WHY WE LAUGH

The Role Of Humour In Human Survival

Dr Bhaskar Bora

DR BHASKAR BORA

A PERSONAL NOTE FROM THE AUTHOR

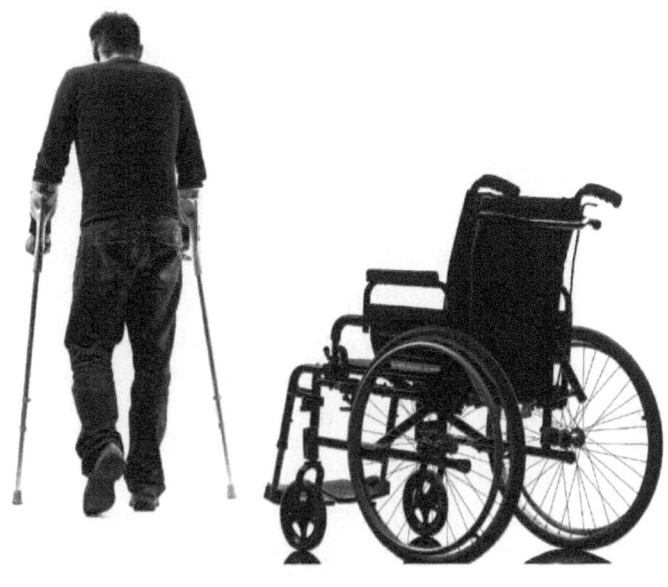

My journey, once marked by certainty and driven by purpose, has transformed in ways I could never have anticipated. It is no longer about grand achievements or the pursuit of external success, but about the quiet, tender moments that reveal the true essence of life—moments of love, care, and presence. What you hold in your hands is not just a collection of words, but a testament to resilience, a story woven from the delicate threads of struggle, acceptance, and ultimately, renewal.

There was a time when my life flowed with the grace of a symphony, every note in perfect harmony. As a doctor, my days were filled with the pulse of life itself—offering hope, easing suffering, and healing with steady hands. The white coat I wore wasn't just a symbol of my profession; it embodied my very identity; an outward reflection of the healer I believed I was destined to be. The lives I touched, the people I helped—it all gave profound meaning to my existence.

But life, in its mysterious and unpredictable ways, had other plans. In one swift, unforeseen moment, the world I knew unravelled. First came the spinal cord injury, stripping away the physical strength I had relied upon. Then, the shadow of cancer darkened the horizon, a stark reminder of life's fragility. The world of medicine, where I once found so much joy and purpose, suddenly slipped away, leaving a vast emptiness in its wake—a silence where once there had been meaning.

Gone were the bustling corridors of the hospital, replaced by the quiet solitude of my home. No longer a "Doctor," I found myself standing at the edge of an uncertain future, my hands—once so steady with the knowledge of healing—trembling with questions I wasn't ready to face. Without the title, without the work that had defined me for so long, who was I? What was left of me when everything I had known was no longer within reach?

In that silence, in the stillness of a life interrupted, I began to uncover something unexpected. The role of a disabled husband and father, once a distant concept, became my new reality—one that held unexpected grace.

What began as an effort to nurture my relationships, to find solace in this new world, slowly evolved into a profound inward journey.

I found healing in the spiritual—a rhythm of meditation, reading, and reflection that allowed me to rediscover the parts of myself I thought were lost. As I immersed myself in books, audiobooks, and hours of research, I began to understand that this new chapter of my life was not an ending, but a rebirth. The solitude of these years, the quiet hours of writing and reflection, gave birth to the very pages you hold in your hands now.

It is with deep gratitude that I share these words with you, knowing that they carry with them not just knowledge, but a piece of my soul. I hope that these reflections and insights offer you a fresh perspective on life and perhaps some nourishment for your own journey.

We cannot control what the universe throws at us, but how we react to those curveballs defines who we are and what we make of our lives.

CHAPTER 1: A LAUGHING MATTER

The Birth Of Humour In Human History

In the deep mists of time, long before humans wielded fire or etched symbols into stone, a sound reverberated through the primaeval forests—a sound as primal and unbidden as the cry of a newborn. It was not a roar of dominance or a whimper of fear, but something lighter, something that danced on the edge of absurdity. It was raw and unrefined laughter bubbling forth from the bellies of creatures who had not yet invented words. This chapter embarks on a journey to uncover the genesis of humour, exploring its humble beginnings among our pre-human ancestors, its role in early human societies, and how it shaped the destiny of an entire species.

The Primal Giggle: Laughter Among Primates
Imagine a troop of chimpanzees in the dense forests of West Africa. Among the group, a young chimp swings wildly on a vine, misjudges the angle, and lands in a heap of leaves. There's a moment of silence as the troop registers the spectacle. Then, as if on cue, a soft panting sound ripples through the group—a rhythmic "ha-ha-ha" that grows louder and more contagious. This is not human laughter, but its unmistakable precursor: the pant-hoot of primates, a sound that evolutionary biologists identify as a "play signal."

Laughter, it turns out, is older than humanity itself. Research by biologist Robert Provine revealed that great apes, particularly chimpanzees, exhibit laughter-like vocalizations during play. These sounds serve as social signals, indicating that rough-and-tumble behaviour is all in good fun, not an act of aggression. The evolutionary purpose? To build trust, diffuse tension, and strengthen social bonds—a theme that persists in human laughter to this day.

But what does this mean for us? If a chimpanzee can laugh, does it understand humour? The answer lies not in the intellect, but in the body. For primates, laughter is a physical response—a release valve for the exuberance of play. It's not the nuanced wit of Oscar Wilde, but it is a foundation: a shared moment of connection in a world where survival often hinges on cooperation.

The Dawn of Human Laughter
Fast-forward several million years to the African savannah, where early humans—our Homo erectus ancestors—are gathering around a communal fire. Fire, a relatively new innovation, has brought light to the night and safety from predators, but it has also created something less tangible: time. Time to relax, to communicate, and to laugh.

These early gatherings likely gave rise to one of the first forms of humour: mimicry. Imagine a hunter returning empty-handed after a failed pursuit, only for a fellow tribe member to imitate his exaggerated gestures and crestfallen expression. The others, weary from the day's toil, burst into laughter. In that moment, a bond is forged, not through words, but through shared amusement.

This laughter served more than a social purpose. It was, in many ways, a survival tool. Humour helped diffuse tensions within the group, reducing the likelihood of conflict. It reinforced hierarchies subtly—leaders who could make others laugh often commanded more respect—and it provided a means of coping with the harsh realities of prehistoric life. A tribe that laughed together survived together, bound by a shared sense of resilience and camaraderie.

The Cave Walls Speak: Humour in Art and Ritual

As humans evolved, so too did their capacity for humour. The earliest evidence of this can be found not in the fossil record, but on the walls of caves. In Chauvet and Lascaux, the famous Palaeolithic cave art sites, depictions of animals and hunting scenes abound, but so too do oddities: a bison with exaggerated features, a figure with a distorted face. Were these merely artistic flourishes, or were they early attempts at humour?

Anthropologists suggest that such images may have served a dual purpose. On one hand, they were didactic tools, teaching young hunters about prey. On the other, they could have been humorous exaggerations meant to entertain and engage. After all, who wouldn't chuckle at the sight of an oversized mammoth tripping over its own feet, even if etched in stone?

Rituals, too, became a fertile ground for humour. Early human ceremonies often involved dancing, music, and mimicry. The shamans of ancient tribes, while revered for their spiritual roles, were also performers, using humour to captivate their audience. This blending of the sacred and the silly underscores a profound truth: humour, even in its earliest forms, was never frivolous. It was a tool for connection, understanding, and survival.

The Evolutionary Theories of Humour

Why, then, did humour evolve? What advantage did it confer to our ancestors that ensured its survival across millennia? Evolutionary scientists have put forth several theories, each shedding light on a different facet of this enigmatic behaviour.

1. The Play Theory: Just as young animals use play to practice survival skills, early humans may have used humour as a form of cognitive play. By engaging in jokes and mimicry, they honed their social intelligence, learning to navigate complex group dynamics.

2. The Social Bonding Theory: Humour, like grooming in primates, served as a way to strengthen social bonds. Unlike grooming, however, laughter could involve multiple individuals simultaneously, making it a more efficient way to build group cohesion.

3. The Sexual Selection Theory: Geoffrey Miller, an evolutionary psychologist, posited that humour evolved as a mating strategy. A sharp wit signalled intelligence and creativity—traits that were highly desirable in a mate. In this sense, laughter wasn't just a social tool; it was an aphrodisiac.

4. The Relief Theory: This theory suggests that humour provides a psychological release, helping early humans cope with stress and fear. By laughing in the face of danger—be it a sabre-toothed tiger or an overbearing group member—they reclaimed a sense of control.

Real-Life Examples: The Laughing Ape to the Laughing Man
Let us pause for a moment to consider the continuity of humour from our primate ancestors to modern humans. Jane Goodall, the renowned primatologist, once observed a group of chimpanzees in Gombe Stream National Park

engaged in a curious game: they took turns pulling each other's tails, emitting panting laughter as they played. The game had no obvious purpose beyond enjoyment, yet it brought the group closer together.

Fast-forward to a modern-day example: the contagious laughter of a child. A toddler giggles uncontrollably at a game of peek-a-boo, a game that, at its core, relies on the same principles of surprise and play. In both cases, laughter transcends language, bridging the gap between individuals through a shared moment of joy.

A Universal Legacy
As we laugh today—at a joke, a meme, or an absurd situation—we are participating in a tradition as old as humanity itself. From the pant-hooting of primates to the mimicry of early humans and the satire of modern comedians, laughter has been a constant companion on our evolutionary journey.

But humour is more than a relic of our past. It is a testament to our resilience, a reminder that even in the harshest conditions, humans found reasons to laugh. And in that laughter, they found strength. After all, to laugh is to affirm life, to proclaim, "I am here, and I am human."

CHAPTER 2: THE SURVIVAL OF THE WITTIEST: HUMOUR IN EVOLUTION

If laughter is the melody of the human spirit, then humour is its secret weapon. Humour and laughter are not mere indulgences—they are evolutionary triumphs that have shaped the course of human survival. From solidifying group bonds to attracting mates, humour has served as a critical tool for navigating the perils and pressures of life on Earth. In this chapter, we explore how humour became a cornerstone of human evolution, focusing on its role in survival, cooperation, and, perhaps most intriguingly, reproduction.

The Social Survival Game
Imagine an early human community, perched on the edge of survival. Resources are scarce, predators loom, and competition within the group itself threatens harmony.

In this precarious world, the ability to navigate complex social interactions could mean the difference between life and death. Enter humour—a seemingly frivolous behaviour with surprisingly potent effects.

Laughter acts as a social lubricant, smoothing over conflicts and fostering group cohesion. In this sense, humour became an evolutionary asset, a way to diffuse tension and build alliances. An early human who could turn a potential argument into a shared chuckle likely stood a better chance of avoiding exile or violence, securing their place in the group. This is where humour's adaptability shines: it can comfort, subvert, or even subtly challenge authority, all without direct confrontation.

Studies in primate behaviour lend credence to this idea. Frans de Waal, a renowned primatologist, observed how bonobos—closely related to humans—use playful behaviours to reconcile after conflicts. A light-hearted tickling game or an exaggerated gesture can signal an end to hostilities. Early humans likely used similar tactics, relying on humour as a non-threatening way to restore harmony.

Laughter in the Wild: Evolutionary Advantages
Beyond its social utility, laughter has physiological benefits that made it a survival-enhancing trait. When early humans laughed, their bodies experienced a cascade of positive effects: stress levels dropped, immune responses were bolstered, and a sense of well-being permeated the group. These benefits were not incidental—they were adaptive.

Consider the scenario of an early hunting party. Returning from an exhausting and dangerous trek, the group faces the disappointment of failure. One member begins to mimic the failed hunt, exaggeratedly sneaking and "tripping" on imaginary roots. The others erupt in laughter. While the food scarcity remains unchanged, the group's spirits lift, rekindling their motivation to try again. Laughter, in this case, serves as a psychological reset, a way to alleviate stress and preserve group morale in the face of adversity.

Modern science has confirmed laughter's stress-relieving function. Studies show that laughter triggers the release of endorphins, the brain's "feel-good" chemicals, and reduces cortisol levels, which are associated with stress. In evolutionary terms, these benefits were life-saving, providing early humans with the resilience needed to endure challenges.

The Sexual Selection Theory: Humour as an Aphrodisiac
Humour's most intriguing evolutionary role, however, lies in the realm of reproduction. Geoffrey Miller, an evolutionary psychologist, posits that humour evolved as a display of intelligence and creativity, traits that were highly desirable in a mate. In his seminal work, The Mating Mind, Miller argues that humour is akin to a peacock's tail—a costly but attractive signal of genetic fitness.

But why would humour be attractive? The answer lies in its complexity. To be funny, one must possess a keen sense of timing, an understanding of social dynamics, and the cognitive flexibility to see the world from novel perspectives. These qualities suggest a robust and adaptable brain, making humour a proxy for overall intelligence. For our ancestors, choosing a mate with a sharp wit was a way to ensure that their offspring would inherit those same advantageous traits.

Real-life studies support Miller's hypothesis. In a 2006 experiment, researchers conducted speed-dating events where participants rated potential mates on various traits. Humour consistently ranked as one of the most attractive qualities, particularly in men. Interestingly, while both sexes appreciated humour, women were more likely to view it as a sign of intelligence and potential compatibility, aligning with the evolutionary theory that humour is a marker of mate quality.

This interplay between humour and attraction is not limited to humans. Studies have shown that male birds, such as bowerbirds, engage in elaborate and playful displays to court females. While not humour in the human sense, these behaviours share a similar function: showcasing creativity and fitness to attract a mate. Humour, then, can be seen as the human equivalent of such displays, a way to dazzle potential partners with cognitive flair.

Real-Life Examples: Wit Wins the Heart
Throughout history, the link between humour and romance has been celebrated. Consider Abraham Lincoln, whose self-deprecating wit and storytelling prowess endeared him to allies and adversaries alike. One of his most famous quips occurred during a political debate, when his opponent accused him of being two-faced.

Lincoln, without missing a beat, replied, "If I had two faces, do you think I'd be wearing this one?" The audience erupted in laughter, cementing his likability and disarming his critic.

Such moments highlight humour's dual power: it can charm while subtly asserting dominance. For Lincoln, humour was not just a personal trait; it was a strategic tool that won him allies and respect, both in politics and in personal relationships.

Modern examples abound as well. Comedians, despite their unconventional careers, are often regarded as highly attractive figures. The allure lies not just in their ability to make others laugh, but in the intellect and charisma required to do so. As a result, humour remains a potent force in courtship, bridging gaps and building connections.

Humour's Role in Parenting
Humour's evolutionary significance extends beyond mate selection to parenting. Parents who can engage their children with humour create stronger emotional bonds and teach resilience. In the unpredictable world of early human life, a playful parent could distract a child from fear or discomfort, turning potential tears into laughter. This emotional regulation was critical for survival, helping children adapt to the stresses of their environment.

Studies in developmental psychology confirm this. Children raised in households where humour is prevalent tend to exhibit higher levels of emotional intelligence and adaptability. These traits, in turn, contribute to their long-term success, reinforcing the evolutionary value of humour as a familial trait.

A Humour-Driven Evolutionary Legacy

Humour's evolutionary advantages are undeniable. It forged social bonds, reduced stress, attracted mates, and strengthened familial ties. More importantly, it ensured the survival of the wittiest—those who could navigate the complexities of human life with a laugh and a smile.

As we laugh today, whether at a clever pun, a viral meme, or the absurdity of our own lives, we are participating in an ancient tradition. Laughter is more than a response; it is an inheritance, a testament to the resilience and creativity of our species. In the words of Victor Borge, "Laughter is the shortest distance between two people." For our ancestors, that distance could mean the difference between survival and extinction.

CHAPTER 3: NEURONS THAT GIGGLE

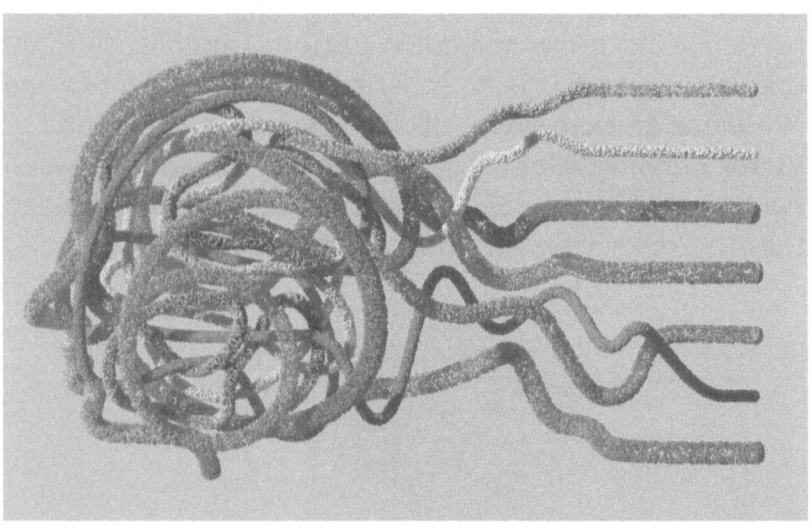

The Neuroscience Of Laughter

It begins with a twitch in the diaphragm, a ripple of sound escaping the lips, and a glimmer in the eyes. Yet behind this seemingly simple act of laughter lies a complex symphony orchestrated by the brain

—a network of neurons firing in harmony, releasing chemical cascades that ripple through the body. Laughter is not just a fleeting expression; it is a profound biological phenomenon. In this chapter, we delve into the neuroscience of laughter, uncovering the brain regions involved, its impact on mental and physical health, and the therapeutic potential of this universal behaviour.

The Brain's Laughter Orchestra
The act of laughing is deceptively complex. Multiple regions of the brain collaborate to produce what feels like a spontaneous reaction. Neuroscientists have identified key areas responsible for generating, perceiving, and responding to humour:

1. The Prefrontal Cortex: This area, particularly the medial prefrontal cortex, is the seat of higher-order thinking. It helps us recognize and process incongruities —the surprising twists and absurdities that make things funny.

2. The Temporal Lobes: The superior temporal gyrus processes auditory input, helping us interpret jokes, wordplay, and comedic timing.

3. The Limbic System: This network, including the amygdala, governs emotional responses. It determines the intensity of our laughter based on the humour's emotional resonance.

4. The Motor Cortex: Once the brain perceives humour, the motor cortex coordinates the physical act of laughing, from the rhythmic contractions of the diaphragm to the facial expressions that accompany it.

Research conducted by Dr Sophie Scott at University College London revealed that laughter activates not just these regions but also the brain's reward circuits. When we laugh, the ventral striatum lights up, releasing a flood of dopamine—a neurotransmitter associated with pleasure and reward.

Laughter's Chemical Cocktail

Laughter is often described as contagious, and its chemistry explains why. When we laugh, the brain releases a potent mix of chemicals that affect both the mind and body:

1. Dopamine: Known as the "feel-good" chemical, dopamine floods the brain's reward pathways, creating a sense of euphoria.

2. Endorphins: These natural painkillers are released during laughter, providing temporary relief from physical discomfort.

3. Oxytocin: Often referred to as the "bonding hormone," oxytocin strengthens social connections, fostering trust and intimacy.

4. Cortisol: Laughter reduces levels of this stress hormone, counteracting the physiological effects of anxiety and tension.

These chemical reactions explain why laughter leaves us feeling relaxed and uplifted. In essence, a hearty laugh is a natural high, a biological gift that promotes both individual well-being and social cohesion.

Mental Health Benefits: A Laughing Mind

The connection between laughter and mental health is profound. Psychologists have long recognized its role in alleviating anxiety, depression, and stress. One landmark study conducted by Dr William Fry, often dubbed the "father of gelotology" (the study of laughter), found that laughter can elevate mood and enhance emotional resilience.

Laughter as a Stress Reliever

When life's burdens grow heavy, laughter acts as a release valve. In one study, participants exposed to humorous stimuli—such as watching comedic videos—experienced a significant drop in cortisol levels. The humour-induced stress reduction persisted long after the laughter subsided, highlighting its enduring effects.

Laughter and Cognitive Function

Interestingly, laughter also enhances cognitive function. Studies have shown that humour activates the brain's creative centres, improving problem-solving skills and mental flexibility. For example, a group of participants who watched a comedy routine before taking a standardized test performed better on questions requiring innovative thinking.

Physical Health Benefits: A Laughing Body

Beyond its mental benefits, laughter has tangible effects on the body. It is, in many ways, a workout disguised as joy. Researchers at Loma Linda University found that laughter improves cardiovascular health, boosts the immune system, and even aids in weight management.

Cardiovascular Health

When you laugh, your heart rate increases briefly, mimicking the effects of moderate exercise. This increase is followed by a period of relaxation, which improves blood flow and reduces blood pressure. In fact, a study published in the journal Heart found that individuals who laughed frequently had a lower risk of heart disease.

Immune System Boost

Laughter also enhances immune function by increasing the production of antibodies and activating T-cells, which fight infections. In one experiment, participants who engaged in a laughter session exhibited higher levels of natural killer cells, which combat viruses and cancer.

Pain Management

Perhaps most remarkably, laughter has been shown to reduce physical pain. This is due to the endorphin release triggered by sustained laughter. Case studies of patients with chronic pain conditions reveal that humour therapy significantly improves their quality of life.

Case Studies: Laughter as Medicine
Norman Cousins: Anatomy of an Illness

One of the most famous anecdotes in laughter therapy involves Norman Cousins, a journalist who suffered from a debilitating autoimmune disease in the 1970s. Cousins, dissatisfied with conventional treatments, devised his own regimen: a combination of vitamin C and laughter. He watched hours of comedic films, including Marx Brothers classics, and reported that ten minutes of belly laughter provided two hours of pain-free sleep. His remarkable recovery inspired further research into the therapeutic power of humour.

Clowns in Hospitals

Building on Cousins' legacy, organizations like Clown Doctors have brought laughter into hospitals worldwide. Paediatric wards, in particular, have embraced humour as a healing tool. A study in Israel found that children undergoing surgery who interacted with medical clowns experienced lower anxiety levels and faster recoveries compared to those who didn't.

Laughter Yoga

Developed by Dr Madan Kataria in the 1990s, laughter yoga combines breathing exercises with voluntary laughter. Practitioners report reduced stress, improved mood, and a sense of community. The practice has spread to over 100 countries, demonstrating the universal appeal of laughter as therapy.

Real-Life Anecdotes: Moments That Heal

In the bustling streets of Mumbai, a laughter club meets at dawn. The members, ranging from young professionals to retirees, begin their day not with meditation but with uninhibited laughter. As the leader shouts, "Ho-ho-ha-ha-ha," the group erupts into synchronized mirth. Strangers transform into companions, and for a brief moment, the weight of the world feels lighter.

In another corner of the world, a cancer patient named Lisa finds solace in a stand-up comedy routine. Despite her gruelling chemotherapy, she laughs until tears stream down her face. "For those few minutes," she says, "I forgot I was sick. I just felt alive."

The Gift of Giggles

Laughter, though fleeting, leaves an indelible mark on the human body and spirit. It connects neurons and people, heals wounds both physical and emotional, and serves as a testament to our resilience. The brain that giggles is a brain that thrives, reminding us that humour is not a mere indulgence but an essential part of our humanity.

As science continues to unravel the mysteries of laughter, one truth stands out: to laugh is to live fully. In the words of the playwright Victor Hugo, "Laughter is the sun that drives winter from the human face." And in its warmth, we find not only joy but also strength, hope, and the promise of a brighter tomorrow.

CHAPTER 4: INSIDE THE LAUGHING BRAIN

Why We Find Things Funny

Humour is a strange and paradoxical phenomenon. A single, well-timed remark can make a room erupt in uncontrollable laughter, while another attempt at levity falls painfully flat. What is it about a joke, a pun, or a simple absurdity that tickles the mind and prompts an outburst of laughter? Understanding why we find things funny requires peeling back the layers of our psychology, exploring theories that explain humour's many dimensions, and weaving in real-life examples that illuminate how humour manifests in everyday moments.

Theories of Humour: The Pillars of Amusement

Psychologists and philosophers have spent centuries dissecting humour, yet its essence often seems elusive. Still, three foundational theories dominate the discussion of why we laugh: superiority, incongruity, and relief. Each theory offers a unique lens through which to view humour and its psychological appeal.

1. The Superiority Theory: Laughing from Above

The oldest explanation of humour, the superiority theory, dates back to Plato and Aristotle. This theory posits that we laugh when we perceive ourselves as better, smarter, or more fortunate than someone else. In essence, humour arises from a sense of triumph or superiority over another's misfortune or folly.

Picture a co-worker slipping on a banana peel in the office hallway. The sudden, exaggerated fall triggers laughter —not necessarily because we're cruel, but because the situation temporarily places us in a superior position. We're not the one who fell, and the absurdity of the moment makes it even funnier.

Modern comedy often draws on this theory. Think of sitcoms where characters like Michael Scott in The Office bumble through social interactions, creating moments where viewers laugh at their awkwardness. The humour isn't always kind, but it reflects a basic human instinct to find joy in comparative advantage.

2. The Incongruity Theory: When Expectations Collide

If the superiority theory feels antiquated, the incongruity theory is its modern counterpart. First articulated by the 18th-century philosopher Immanuel Kant, this theory suggests that humour arises when there's a mismatch between expectation and reality. The punchline of a joke, for instance, surprises us by taking an unexpected turn, creating cognitive dissonance that resolves in laughter.

Imagine you're walking your dog and see a sign that reads, "Lost Dog: Answers to 'Cat.'" The absurdity of the statement catches you off guard, and before you can process it, you're laughing. The humour lies in the incongruity—the mismatch between the expectation of a normal lost dog sign and the surprising detail of the name "Cat."

This theory also explains why timing is crucial in comedy. A delayed punchline or a well-placed pause enhances the incongruity, making the surprise more impactful. Stand-up comedians like Mitch Hedberg and Demetri Martin excel in this domain, delivering observations that twist mundane situations into hilarious incongruities.

3. The Relief Theory: Humour as Emotional Release
The relief theory, championed by Sigmund Freud, takes a different approach. Freud believed that humour acts as a pressure valve, releasing pent-up energy or tension. Whether we're laughing at a taboo topic, a stressful situation, or an embarrassing moment, humour provides psychological relief.

You're in a tense meeting, and the atmosphere is suffocatingly serious. Suddenly, someone makes a self-deprecating remark about forgetting their notes. The room bursts into laughter, and the tension dissipates. Here, humour serves as a coping mechanism, allowing everyone to relax and reset.

Relief theory also explains why people often laugh at dark humour or jokes about topics that are otherwise uncomfortable. It's not that they find the subject matter inherently funny, but the humour provides a way to navigate discomfort, fear, or sadness. Shows like BoJack Horseman and Rick and Morty leverage this tension, blending humour with existential themes to create cathartic laughter.

Everyday Humour: Psychology at Play
Understanding these theories helps decode the humour in everyday life, where ordinary moments often turn unexpectedly funny. Let's explore a few scenarios and analyse their psychological underpinnings:

Scenario 1: The Meeting Mishap
During a Zoom call, an executive forgets to unmute before launching into a long-winded explanation. The team sits silently, watching his animated gestures while hearing nothing. When someone finally points out the mistake, laughter erupts.

- Theories at play: Incongruity (unexpected silence on a video call) and superiority (recognizing that it's not us making the mistake).

Scenario 2: The Freudian Slip
At a wedding, a nervous best man accidentally says, "We're all here to celebrate the loving union of Jack and Jill—oops, I mean Jane!" The room bursts into laughter at the unintended error.

- Theories at play: Relief (breaking the tension of a formal event) and incongruity (the mismatch between expectation and the slip).

Scenario 3: The Parent's Pun

A child spills milk at the breakfast table, and the father quips, "Well, no use crying over it now!" The groan-laugh combo from the family reflects the universal appeal (and irritation) of puns.

- Theories at play: Incongruity (wordplay creating a mismatch) and relief (diffusing potential frustration over the spill).

Cultural Variations in Humour
Humour isn't universal in its delivery or interpretation. What elicits laughter in one culture might fall flat—or even offend—in another. The psychological theories of humour remain consistent, but their applications vary across cultural contexts.

- Superiority in Culture: In British comedy, self-deprecating humour is a staple, inviting audiences to laugh with, rather than at, the comedian. This contrasts with the more assertive humour often found in American stand-up.

- Incongruity in Culture: Japanese humour often relies on boke and tsukkomi routines, where one performer says something absurd (boke), and the other reacts with a sharp, corrective remark (tsukkomi). This interplay of incongruity resonates deeply with Japanese audiences.

- Relief in Culture: In India, political satire provides a much-needed outlet for navigating societal tensions. Shows like The Kapil Sharma Show use humour to address sensitive topics in ways that feel safe and communal.

Humour Across Life Stages

Humour isn't static; it evolves as we age. Children, for instance, delight in slapstick and physical comedy, while adults gravitate toward more nuanced humour that reflects their experiences and intellect.

- Children and Humour: A toddler laughs at peek-a-boo because the sudden reappearance of a hidden face creates a predictable but delightful incongruity.

- Teenagers and Humour: Adolescents often use humour to navigate social hierarchies, with sarcasm and irony becoming prominent tools.

- Adults and Humour: For adults, humour often blends intellectual and emotional elements, reflecting life's complexities. A clever pun or a dark joke might elicit laughter, depending on context and personality.

Laughter, the Mirror of the Mind

In the end, why we find things funny is as much about psychology as it is about the human experience. Humour reflects our ability to think abstractly, cope with stress, and navigate social dynamics. It is both a mirror of the mind and a bridge between individuals, connecting us through shared amusement.

As we laugh at the absurdities of life, we are participating in something deeply human. Laughter reminds us that, despite our struggles and imperfections, there is always room for joy, surprise, and connection. In the words of Charlie Chaplin, "A day without laughter is a day wasted."

CHAPTER 5: THE HEALING POWER OF HUMOUR

In the darkest corners of human experience, where despair looms large and hope feels out of reach, humour emerges as a surprising yet steadfast companion. Laughter, at its core, is an act of defiance against suffering. It signals resilience, a refusal to surrender to the weight of trauma and stress. In this chapter, we explore how humour becomes a lifeline in adversity, examining studies, including Viktor Frankl's observations in concentration camps, and recounting real-life stories of individuals who have turned to humour to navigate their most challenging moments.

Humour as a Coping Mechanism
When life delivers its heaviest blows, the instinct to laugh might seem counterintuitive. Yet, psychologists have found that humour plays a critical role in how people process and endure trauma. By reframing painful experiences through humour, individuals can achieve emotional distance, reduce anxiety, and reclaim a sense of control over circumstances that feel otherwise

uncontrollable.

The Role of Reframing

Humour enables a process psychologists call "reframing"—the act of viewing a distressing situation from a new, less threatening perspective. For instance, a woman who accidentally spilled coffee on her wedding dress might quip, "Well, at least it's not red wine!" While the mishap remains, the humour changes how it's perceived, making it more manageable.

Psychological Resilience

Studies have consistently shown that humour enhances psychological resilience. People who incorporate humour into their coping strategies tend to experience lower levels of stress and depression. Laughter doesn't erase the pain, but it acts as a buffer, allowing individuals to confront challenges with renewed strength.

Viktor Frankl: Laughter Amid the Unthinkable

One of the most profound explorations of humour in the face of trauma comes from Viktor Frankl, an Austrian psychiatrist and Holocaust survivor. In his seminal work, Man's Search for Meaning, Frankl recounts how humour helped him and his fellow prisoners survive the unimaginable horrors of Nazi concentration camps.

Frankl described how he and another prisoner invented a "joke of the day" routine, trading quips about their dire circumstances. They found ways to laugh at the absurdity of their situation—not to diminish its gravity, but to momentarily transcend it. "Humour was another of the soul's weapons in the fight for self-preservation," Frankl wrote. "It is well known that humour, more than anything else in the human makeup, can afford an aloofness and an ability to rise above any situation, even if only for a few seconds."

This use of humour wasn't about ignoring reality. It was about reclaiming agency in a place designed to strip it away. By laughing, Frankl and his peers reaffirmed their humanity in a setting that sought to erase it.

Real-Life Stories: Humour in Hardship
The power of humour to heal and endure extends beyond concentration camps. Time and again, individuals across the world have turned to laughter to navigate personal crises. Here are true accounts of how humour has served as a lifeline in adversity.

1. The Cancer Patient Who Wrote Her Own Obituary
Valerie Harper, an actress best known for her role on The Mary Tyler Moore Show, faced terminal cancer with remarkable humour. In interviews, she joked about her prognosis, saying, "We're all terminal; I just have a better sense of when." Her candid humour about her mortality helped her friends and family process her illness with less fear and more openness.

Similarly, Annie Schmidt, diagnosed with stage-four breast cancer, wrote her own obituary, filling it with jokes and wit. She described herself as "a mediocre cook but a great eater" and ended with the line, "On the bright side, I won't have to worry about receding gums or bunions anymore." Schmidt's humour helped her loved ones remember her with joy rather than sorrow.

2. The Firefighter's Joke

In 2017, firefighter Mike Hughes sustained severe burns rescuing people from a wildfire in California. During his painful recovery, he turned to humour to cope with the overwhelming physical and emotional challenges. When asked about his injuries, he quipped, "At least I don't have to shave anymore!" His self-deprecating jokes became a way to connect with others and diffuse the heavy emotions surrounding his situation.

3. The Soldier's Laugh
During the Gulf War, British soldier Stephen Brown found himself pinned down under heavy fire. With death seeming imminent, a fellow soldier made a sarcastic comment about their predicament: "Well, this is a fine mess you've got us into!" The group burst into laughter, momentarily breaking the tension and allowing them to regroup. Brown later reflected, "That laughter saved us. It gave us just enough clarity to keep going."

The Science of Humour in Healing
The benefits of humour during trauma are not merely anecdotal; they are supported by robust scientific evidence. Researchers have identified several mechanisms by which humour alleviates stress and aids in recovery:

1. Reduction of Stress Hormones
Laughter reduces cortisol, the hormone associated with stress, while increasing endorphins, which promote a sense of well-being. Studies have shown that people who laugh frequently exhibit better stress management and emotional regulation.

2. Improved Social Connections

Shared laughter strengthens social bonds, creating a sense of solidarity and mutual support. In group settings, such as hospitals or shelters, humour fosters a sense of community, reducing feelings of isolation and despair.

3. Physical Benefits
Humour has tangible effects on the body. Laughter boosts immune function, lowers blood pressure, and increases oxygen intake, all of which contribute to improved physical health. In rehabilitation centres, laughter therapy sessions have led to faster recoveries and better patient outcomes.

Communities That Laugh Together
The healing power of humour extends to entire communities. Following the devastation of Hurricane Katrina, New Orleans residents turned to their unique brand of humour to rebuild their spirits. Local artists painted satirical signs mocking the slow government response, and Mardi Gras parades featured floats poking fun at the disaster. These acts of humour didn't erase the tragedy, but they provided a way for the community to process their grief collectively.

Similarly, during the COVID-19 pandemic, humour became a global coping mechanism. Memes about Zoom calls, toilet paper shortages, and sourdough bread dominated social media, providing a sense of connection during isolation. Laughter, even in the face of a global crisis, reminded people that they were not alone.

Laughter as a Lifeline
In the face of trauma and stress, humour serves as a

bridge between despair and hope. It doesn't erase pain but transforms it, offering a new perspective and a momentary reprieve. The ability to laugh in adversity is not a denial of reality—it is an affirmation of humanity's enduring spirit.

Whether it's Viktor Frankl cracking jokes in a concentration camp, a firefighter finding levity in recovery, or communities rallying through satire, humour proves time and again that it is not merely entertainment. It is survival. As Charlie Chaplin famously said, "To truly laugh, you must be able to take your pain and play with it." In that playfulness, we find strength, resilience, and the will to carry on.

CHAPTER 6: THE DARK SIDE OF THE JOKE

When Humour Hurts

Humour, often celebrated as a force for connection and healing, has a shadow side. Like a double-edged sword, it can foster joy and understanding but also wound, alienate, or manipulate. In the wrong context or with malicious intent, humour becomes a weapon, wielded to diminish others or assert control. This chapter examines the darker aspects of humour, exploring how it can be divisive, cruel, and even dangerous, supported by real-life examples and studies on its dual nature in social dynamics.

The Alienating Power of Humour

Laughter often creates a bond between those who share it, but it can also exclude and ostracize. Inside jokes, for example, strengthen ties among those in the know while alienating outsiders. This dynamic plays out in countless social settings, from playgrounds to boardrooms, where humour becomes a subtle yet potent tool for defining who belongs—and who doesn't.

Example: The Office Cliques
In a corporate setting, humour can serve as a gatekeeper for group inclusion. Imagine a workplace where a small team shares jokes steeped in sarcasm and cultural references. For new hires or employees unfamiliar with the humour, the laughter feels like a closed door, reinforcing a sense of "us versus them." Instead of fostering camaraderie, humour here becomes a wall.

Studies in organizational behaviour highlight this phenomenon. Researchers have found that exclusive humour in professional environments creates in-groups and out-groups, often marginalizing those not fluent in the group's comedic language. What begins as light-hearted banter can evolve into a barrier to diversity and inclusivity.

Humour as a Weapon: Insult and Bullying

Some of the darkest uses of humour occur when it becomes a deliberate tool for insult or humiliation. While it might masquerade as a joke, hurtful humour often carries the sting of aggression wrapped in laughter, leaving its target disarmed and unable to retaliate without appearing humourless or overly sensitive.

Example: The Bullying Mask
Consider a high school where a student, Alex, is ridiculed for their appearance. A classmate might jokingly comment, "Nice outfit—straight out of the 80s, right?"

The remark draws laughs from the crowd, but for Alex, it cuts deep. Because it's framed as humour, the perpetrator can deflect responsibility, claiming, "It's just a joke!" Meanwhile, the humiliation lingers, eroding Alex's confidence.

This dynamic extends into adult interactions as well. Workplace harassment, for instance, often cloaks itself in humour. A manager might make offhand jokes about an employee's accent or mannerisms, framing it as playful banter. The laughter in the room disguises the underlying hostility, creating an environment where the victim feels isolated and powerless.

Political Satire: A Fine Line Between Critique and Insult

Political satire exemplifies humour's dual nature. At its best, satire challenges authority, exposes hypocrisy, and fosters critical thinking. At its worst, it devolves into personal attacks, alienating those it seeks to engage.

Example: The Charlie Hebdo Controversy
The French satirical magazine Charlie Hebdo is both celebrated and criticized for its provocative humour. In 2015, the magazine published cartoons mocking religious figures, sparking outrage and violent retaliation. Supporters hailed the magazine as a bastion of free speech, while critics argued that its humour crossed into cultural insensitivity and incitement.

This incident underscores the delicate balance of satire. While it aims to hold power to account, it can also deepen societal divides, particularly when the humour targets deeply held beliefs or marginalized groups.

Studies on satire reveal this tension. Research from the University of Michigan found that political humour often reinforces preexisting biases. People tend to laugh more at jokes that align with their views and reject humour that challenges them, making satire a polarizing rather than unifying force.

Dark Humour: Coping or Cruelty?

Dark humour, which confronts taboo subjects like death, tragedy, and suffering, occupies a morally ambiguous space. For some, it serves as a coping mechanism, allowing them to navigate difficult emotions. For others, it feels insensitive or even cruel.

Example: Gallows Humour in Medicine
In hospitals, doctors and nurses often use dark humour to cope with the emotional toll of their work. A surgeon might jokingly say, "At least we won't have to worry about malpractice—this patient's already a goner," after losing a critical case. Among peers, the humour provides relief, a way to process grief and frustration. But to an outsider, such remarks might appear callous or disrespectful.

Example: Online Memes
During the COVID-19 pandemic, dark humour proliferated on social media. Memes about hoarding toilet paper, Zoom fatigue, and even mortality became viral hits. For some, these jokes offered a sense of solidarity in shared hardship. For others, they trivialized the severity of the crisis, sparking debates about whether such humour was appropriate.

Psychologists have found that individuals with higher

emotional resilience are more likely to appreciate dark humour. A study published in Cognitive Processing found that people who enjoyed dark jokes scored higher on measures of intelligence and emotional stability. However, the same humour that provides catharsis for one person might alienate or offend another, highlighting its subjective nature.

Humour in Manipulation and Propaganda

Humour's persuasive power makes it a tool not only for individuals but also for institutions and governments. Throughout history, propaganda has used humour to ridicule opponents, dehumanize enemies, and rally support for causes.

Example: WWII Propaganda Cartoons
During World War II, Allied and Axis powers alike produced cartoons mocking their adversaries. The caricatures reduced complex political and moral conflicts to simplistic, often racist, depictions, fuelling animosity and justifying violence. While the humour united citizens under a common cause, it also perpetuated harmful stereotypes and escalated division.

Example: Modern Political Memes
In the digital age, memes have become a weapon in ideological battles. Political campaigns and activists use humorous images and videos to sway public opinion. While some memes foster engagement, others rely on misinformation and ridicule, polarizing audiences and spreading false narratives.

Studies: The Dual Nature of Humour in Social Dynamics

The dual nature of humour—its ability to connect and divide—is well-documented in psychological and sociological research. A study by Dr Rod Martin, a leading humour researcher, identified four distinct styles of humour:

1. Affiliative Humour: Builds bonds and fosters connection.
2. Self-Enhancing Humour: Helps individuals cope with stress and adversity.
3. Aggressive Humour: Targets others through insults and sarcasm.
4. Self-Defeating Humour: Involves making oneself the butt of the joke to gain approval.

While affiliative and self-enhancing humour tend to have positive effects, aggressive and self-defeating humour often create negative social outcomes, including alienation and conflict.

The Responsibility of Humour

Humour is neither inherently good nor bad; it is a tool, its impact shaped by intent, context, and perception. While it can bridge divides and bring joy, it also has the power to wound, manipulate, and alienate. Recognizing humour's dual nature is essential to wielding it responsibly.

As we laugh at a joke or share a meme, we must ask ourselves: Who benefits from this humour, and who might it harm? In doing so, we can ensure that our laughter uplifts rather than diminishes, connects rather than divides. After all, as Mark Twain once said, "The secret source of humour is not joy but sorrow. There is no humour in heaven."

CHAPTER 7: FROM CAVE PAINTINGS TO MEMES

Humour Across Eras

Humour is as old as humanity itself, an ever-evolving reflection of culture, creativity, and collective experience.

From the mischievous scribbles on cave walls to the fleeting laughs of internet memes, humour has travelled through art, literature, theatre, film, and digital culture, shaping and mirroring society at every turn. This chapter traces the journey of humour across eras, highlighting pivotal moments and figures that defined its transformation, and examines what our laughter tells us about who we are.

The Earliest Laughs: Cave Paintings and Primitive Play
Humour's origins are as primal as survival itself. Even before the advent of written language, humans expressed their wit through images and gestures. Cave paintings, long viewed solely as ritualistic or instructional, have recently revealed another layer: the comedic.

Example: Exaggerated Animal Depictions
In Chauvet and Lascaux, Palaeolithic cave art depicts not only majestic hunting scenes but also humorous exaggerations—bison with overly large horns or human-animal hybrids in absurd poses. Anthropologists speculate these depictions were intentional, meant to amuse or teach through humour. A hunter drawing an oversized mammoth could evoke laughter while emphasizing caution: "Don't mess with something this big!"

This interplay of art and humour wasn't limited to drawings. Early humans likely used physical comedy—mimicry, exaggerated gestures, and slapstick—to entertain around the fire. Their laughter was both a bonding mechanism and a stress reliever, grounding them in shared humanity amidst the harsh realities of prehistoric life.

The Wit of Words: Humour in Classical Literature

With the rise of written language, humour expanded its repertoire. Ancient civilizations infused their literature with satire, irony, and playful storytelling, offering a glimpse into their values and struggles.

Example: Aristophanes' Satirical Genius

In 5th-century BCE Athens, Aristophanes dominated the comedic stage with plays like Lysistrata, which humorously depicted women withholding sex to end a war. Through sharp satire, Aristophanes critiqued politics, gender roles, and societal norms. His works remind us that humour has long been a tool for challenging authority and sparking dialogue.

Example: Indian Subcontinent's Sanskrit Humour

In ancient India, Sanskrit plays like Mṛcchakatika (The Little Clay Cart) incorporated humour through clever dialogue, mistaken identities, and witty servants. These comedies balanced serious themes with levity, showcasing how humour provided both entertainment and reflection.

The classical period established humour as a vital narrative force, laying the groundwork for literary giants who would elevate it to new heights.

Renaissance Revelry: Shakespeare's Timeless Humour

No discussion of humour's evolution would be complete without William Shakespeare. The Bard's mastery of wit, wordplay, and comedic timing cemented his place as a pillar of literary humour. His comedies, such as *A Midsummer Night's Dream* and *Much Ado About Nothing*, combined physical gags, clever banter, and complex characters to explore love, identity, and human folly.

Example: The Quick-Tongued Fool
In *Twelfth Night*, the character Feste embodies Shakespeare's genius for humour. As a court jester, Feste wields his wit not just for laughs but to reveal uncomfortable truths. When he quips, "Better a witty fool than a foolish wit," his humour transcends mere entertainment, challenging the audience to question societal pretensions.

Shakespeare's humour endures because it strikes a delicate balance between universality and specificity, weaving timeless truths with the quirks of Elizabethan England.

Silent Laughter: The Birth of Film Comedy
The 20th century heralded a new era for humour with the advent of cinema, and silent films became its first stage. Without spoken dialogue, comedians relied on physicality, timing, and visual absurdity to provoke laughter.

Example: Charlie Chaplin's Everyman
Charlie Chaplin's "Little Tramp" character epitomized humour's ability to blend comedy with commentary. In *Modern Times* (1936), Chaplin hilariously struggles with industrial machinery, poking fun at the dehumanizing effects of modernity. His slapstick routines—tripping over obstacles, juggling impossible tasks—were

universally relatable, transcending language and cultural barriers.

Chaplin's humour wasn't just escapism; it was empathy. Through laughter, he exposed the absurdities of societal change while reminding audiences of their shared struggles and humanity.

Satire on Stage: The Power of Stand-Up Comedy

As film and television evolved, so too did the stage. Stand-up comedy emerged as a medium where individuals could wield humour as a weapon for reflection, rebellion, and connection.

Example: Richard Pryor's Raw Truth

In the 1970s, Richard Pryor redefined stand-up comedy by blending humour with vulnerability. Through jokes about race, poverty, and addiction, Pryor forced audiences to confront uncomfortable truths while laughing at their absurdity. His humour was a mirror, reflecting societal issues while forging an intimate bond with his audience.

Stand-up comedy underscored humour's power to amplify marginalized voices, challenging the status quo one punchline at a time.

The Digital Revolution: Memes and Viral Humour

In the 21st century, humour entered a new frontier: the internet. Memes, GIFs, and viral videos became the dominant comedic language, spreading laughter across the globe in seconds.

Example: The Distracted Boyfriend Meme

One of the most recognizable internet memes features a man turning to ogle a passing woman while his girlfriend glares at him. This single image, repurposed with countless captions, encapsulates humour's modern evolution. It relies on incongruity—juxtaposing a familiar scenario with absurd commentary—to elicit laughter.

Example: TikTok Comedy
Platforms like TikTok have democratized humour, allowing anyone with a smartphone to create and share jokes. From lip-syncing skits to surreal edits, TikTok humour often thrives on relatability and absurdity. During the COVID-19 pandemic, for instance, creators turned everyday frustrations—like Zoom call mishaps—into comedic gold, offering a shared sense of relief in isolation.

Digital humour is fleeting yet impactful, capturing the zeitgeist while reflecting humour's timeless ability to connect and adapt.

The Universality and Specificity of Humour
Across eras, humour has been both universal and specific. It transcends language and geography, yet it remains deeply rooted in the culture and context of its time. Shakespeare's witty repartee resonates today, but only because his themes—love, identity, human folly—are timeless. Similarly, a meme about internet culture might provoke laughter now but fade into obscurity as technology evolves.

What remains constant is humour's ability to reflect humanity's complexities, bridging the past and present in a way few other art forms can.

Laughter Through Time

From cave paintings to memes, humour has evolved alongside humanity, adapting to new mediums while retaining its core essence: connection, commentary, and catharsis. Each era's humour is a snapshot of its people, their struggles, and their joys. Whether we're laughing at Charlie Chaplin's antics or a TikTok skit, we are participating in an age-old tradition, one that unites us across time and space.

In humour, we find not just entertainment but a mirror, a bridge, and a balm. As we laugh at the absurdities of life, we carry forward the legacy of those who laughed before us, ensuring that humour remains a vital thread in the tapestry of human experience.

CHAPTER 8: HUMOUR AS A SOCIAL GLUE

The Bonding Power Of Laughter

Laughter isn't just an expression of joy—it's a declaration of belonging. It breaks barriers, builds bridges, and binds people together in ways few other behaviours can. Humour is more than entertainment; it is a vital social tool, deeply embedded in our evolutionary and cultural history. Whether among friends, teams, or entire communities, humour creates a shared language that fosters trust, solidarity, and understanding.

This chapter dives deeper into humour's role as a social glue, exploring how it connects individuals and groups across various contexts. From its biological foundations to its modern manifestations, we'll examine studies, real-life examples, and cultural nuances to reveal the transformative power of shared laughter.

The Evolutionary Role of Humour in Social Bonding

Laughter is a primal act. Long before humans could articulate words, laughter signalled safety and connection. Robin Dunbar's research suggests that laughter evolved as a social bonding mechanism, playing a role akin to grooming in primates. While grooming is limited to one-on-one interactions, laughter could engage an entire group simultaneously, making it a more efficient way to maintain social cohesion.

The Biochemistry of Connection
When we laugh, our brains release endorphins, chemicals that reduce stress and create a sense of pleasure. This biochemical reaction doesn't just make us feel good—it strengthens our emotional bonds with those we're laughing with. Dunbar's studies reveal that shared laughter increases pain tolerance, a physiological marker of heightened endorphin activity, and fosters feelings of closeness.

Why Laughter is Contagious

Humour is inherently social because laughter is contagious. MRI studies show that hearing laughter activates regions of the brain associated with social responses. In a sense, our brains are wired to respond to laughter as an invitation to connect. This is why a single person's chuckle in a quiet room can spark a chain reaction of laughter, drawing people into an unspoken agreement of camaraderie.

Humour and Friendship: The Foundation of Connection

Laughter is often the first sign of a budding friendship. Shared humour reveals compatibility, creates rapport, and lays the groundwork for deeper connections. When we laugh with someone, we're signalling that we understand and accept their perspective, fostering a sense of mutual trust.

Example: The Coffee Spill Friendship

Take Emily and Sarah, who met during a stressful group project in college. One day, Emily accidentally spilled coffee all over her notes. While others gasped, Sarah burst into laughter and joked, "Well, there goes our chances of graduating." Emily laughed too, and that shared moment of levity broke the ice. Their bond, forged in humour, grew into a lifelong friendship.

Studies reinforce the connection between shared laughter and closeness. Research published in Behavioural and Brain Sciences found that friends who laugh together report higher levels of trust and intimacy than those who don't. Humour acts as a litmus test for compatibility, helping people decide whether a potential friendship will thrive.

The Role of Humour in Teams: Productivity Through Play

In group settings, humour transforms individuals into cohesive units. Teams that laugh together not only perform better but also navigate challenges more effectively. Shared humour creates an environment of psychological safety, encouraging team members to take risks, share ideas, and collaborate without fear of judgment.

Example: "Laughing Mondays" at the Startup
At a Silicon Valley software startup, the CEO implemented "Laughing Mondays," where employees began the week with a short session of comedic storytelling or meme-sharing. Initially, it seemed like a frivolous exercise. But within months, the workplace culture shifted. Collaboration increased, conflicts became easier to resolve, and employee satisfaction soared. "Humour broke down barriers," the CEO observed. "It reminded us we're all human first."

A 2020 study in Journal of Business and Psychology echoes this experience, showing that workplace humour enhances team cohesion and creativity. However, it also cautions against sarcasm or targeted jokes, which can have the opposite effect by creating divisions.

Why It Works
Humour lightens the emotional load of group interactions. A tense brainstorming session can become productive when someone makes a well-timed joke,

breaking the tension and resetting the group's focus. Moreover, humour humanizes leaders, making them more approachable and fostering trust among team members.

Humour in Communities: Creating a Shared Identity

Humour doesn't just connect individuals—it unites entire communities. Shared laughter becomes a collective language, a way of defining "us" versus "them" while reinforcing a sense of belonging. Whether in the aftermath of a disaster or during a festival, humour plays a pivotal role in building communal resilience and identity.

Example: Post-Hurricane Katrina in New Orleans
After Hurricane Katrina devastated New Orleans, humour became a coping mechanism and a rallying cry. Mardi Gras floats featured satirical depictions of the government's slow response, while local comedians turned the chaos of evacuation into punchlines. One popular joke went: "I put FEMA on speed dial—number 404. Not found." This humour didn't erase the pain, but it allowed residents to process their grief collectively, turning tragedy into a shared narrative of resilience.

Sports Fans and Humour
Humour thrives in sports fandom, where jokes, chants, and memes become expressions of loyalty and rivalry. A clever chant at a soccer match—mocking the opposing team's recent loss—can energize a crowd and solidify a sense of camaraderie among fans. These humorous expressions are not trivial; they are rituals that reinforce collective identity.

Cultural Perspectives on Humour and Connection

While humour is universal, its expressions and interpretations are shaped by culture. In Japan, humour often relies on boke and tsukkomi routines, where one person plays the fool (boke) and the other reacts with a witty retort (tsukkomi). This dynamic emphasizes subtlety and timing, reflecting Japan's cultural preference for harmony and nuance.

In contrast, British humour often leans on sarcasm and self-deprecation, allowing people to bond by poking fun at themselves or societal norms. Meanwhile, American humour tends to favour boldness and relatability, emphasizing comedic storytelling and larger-than-life characters.

Example: Cross-Cultural Humour in Diplomacy
During a diplomatic meeting, a British delegate opened with a self-deprecating quip: "I tried to cook last night. Let's just say the fire department now knows my name." The Japanese delegation responded with polite laughter, sharing their own anecdotes about failed cooking attempts. This simple exchange, grounded in humour, broke the ice and fostered goodwill between the two groups.

Digital Humour: The New Frontier of Social Connection

In the digital age, humour transcends physical boundaries, connecting people across the globe. Memes, GIFs, and viral videos have become the lingua franca of the internet, creating virtual communities bonded by shared laughter.

Example: Pandemic Memes
During the COVID-19 pandemic, memes about Zoom calls, toilet paper shortages, and sourdough baking became sources of shared amusement. While the world faced unprecedented isolation, these digital jokes reminded people they were not alone. Humour, even in a virtual space, created a sense of solidarity.

Humour Gone Wrong: When Laughter Divides
Not all humour fosters connection. Misplaced or exclusionary jokes can alienate and harm. A poorly timed joke in a wedding toast or a sarcastic remark in a workplace can create divisions rather than unity. Understanding the context and audience is essential to ensuring humour remains inclusive and constructive.

Example: The Off-Colour Wedding Speech
At a wedding, a best man joked about the groom's embarrassing teenage habits. While some guests laughed, others felt uncomfortable, including the bride's family. Instead of creating joy, the joke underscored cultural differences, highlighting the need for humour to consider context and shared values.

Laughter as the Universal Language

From the intimate laughter of friends to the collective joy of communities, humour binds us in profound and enduring ways. It transcends words, bridging divides and fostering understanding. In a world often marked by division, laughter reminds us of our shared humanity.

As Robin Dunbar's research reveals, laughter is more than a reflex—it is a social superpower. Whether bonding over a joke in a boardroom or sharing memes across continents, humour remains one of our most powerful tools for connection. As we laugh together, we affirm that, despite our differences, we are all part of the same human story.

In the words of Maya Angelou: "I don't trust anyone who doesn't laugh." And perhaps, in that shared laughter, we find the trust and unity we need to face the complexities of life—together.

CHAPTER 9: LEADERS WHO LAUGH

Humour In Power And Politics

Laughter is a universal language, and in the realm of power and politics, it serves as more than just a tool for amusement. For leaders, humour is a strategic weapon, capable of breaking tension, building trust, and persuading sceptics. It humanizes authority, making leaders appear relatable while also disarming criticism and rallying support. From the self-deprecating wit of Abraham Lincoln to the razor-sharp humour of Winston Churchill and the meme-savvy leaders of today, humour has consistently proven to be one of leadership's most potent, yet underestimated, tools.

This chapter takes a deeper dive into how leaders have wielded humour effectively across history and explores its modern-day applications, while also acknowledging its risks.

The Power of Humour in Leadership

Leadership demands the ability to inspire, connect, and influence. Humour uniquely facilitates all three. It lightens the burden of serious issues, strengthens group cohesion, and enables leaders to address contentious topics without alienating their audience. Psychologists and leadership experts emphasize the role of humour in signalling confidence, emotional intelligence, and approachability—all traits critical for effective leadership.

How Humour Works in Leadership
1. Humanizing Authority: Humour bridges the gap between leaders and their followers, making even the most powerful figures appear relatable.
2. Defusing Tension: A well-timed joke can neutralize hostility, whether in a heated debate or a diplomatic negotiation.
3. Amplifying Messages: Humour makes messages more memorable and persuasive, ensuring they resonate with the audience long after delivery.
4. Creating Resilience: Leaders who use humour signal optimism and strength, especially during crises, inspiring followers to remain steadfast.

Abraham Lincoln: The Self-Deprecating Storyteller

Few leaders in history have harnessed the power of humour as effectively as Abraham Lincoln. His presidency during the Civil War was marked by immense strain, yet Lincoln often turned to humour to connect with the American people, ease tensions, and navigate difficult conversations.

Example: Turning Insults Into Laughter
One of Lincoln's most famous displays of humour occurred during a debate when his opponent accused him of being two-faced. Without hesitation, Lincoln quipped, "If I had two faces, would I be wearing this one?" The audience erupted in laughter, and Lincoln's wit neutralized the attack, turning the insult into a moment of charm.

Example: Using Humour to Build Resilience
In the depths of the Civil War, Lincoln would often lighten his Cabinet meetings with humorous anecdotes. He once remarked, "I laugh because I must not cry. That is all. That is all." This acknowledgment of humour as a coping mechanism resonated with those around him, fostering a sense of unity and determination in the face of overwhelming adversity.

Lincoln's humour wasn't just for entertainment—it was a deliberate strategy to connect with both allies and opponents, bridging divides with a shared sense of humanity.

Winston Churchill: Wit as a Weapon

Winston Churchill's wit was as sharp as his strategic mind. During World War II, Churchill used humour not only to rally his nation but also to navigate the treacherous waters of international diplomacy.

Example: Sparring with Lady Astor
One of Churchill's most famous exchanges occurred with

Lady Astor, a political rival. When she said, "If I were your wife, I'd poison your tea," Churchill retorted, "Madam, if I were your husband, I'd drink it." This quick-witted reply showcased his ability to deflect hostility with humour, earning him admiration even from his detractors.

Example: Inspiring a Nation
Churchill's speeches during the Blitz were filled with humour that inspired resilience. After a particularly devastating air raid, he quipped, "You can always count on Americans to do the right thing—after they've tried everything else." While humorous, the remark highlighted his pragmatism and ability to maintain perspective under pressure.

Churchill's humour was not frivolous; it was strategic. It rallied his people during their darkest hours, building morale and reinforcing a collective sense of purpose.

Modern Leaders and Humour in the Digital Age

In today's interconnected world, humour has taken on new dimensions. Leaders now use humour not just in speeches but also on social media and in viral content to connect with younger audiences and navigate the fast-paced demands of public discourse.

Barack Obama: Humour as Relatability
Barack Obama's use of humour made him one of the most relatable modern leaders. Whether delivering zingers at the White House Correspondents' Dinner or reading "mean tweets" on Jimmy Kimmel Live, Obama used humour to build a connection with audiences across political divides.

Example: At a correspondents' dinner, Obama joked about being accused of socialism, saying, "Why don't you let me be clear: My next plan for America involves a little something I like to call 'Obamacare for cats.'" This playful exaggeration not only entertained but also diffused tension around a polarizing policy.

Obama's humour reflected his confidence and composure, making him approachable while subtly reinforcing his authority.

Volodymyr Zelenskyy: The Comedian-Turned-President
Ukrainian President Volodymyr Zelenskyy's background as a comedian gives him a unique edge in political leadership. During his campaign, his humour resonated with citizens disillusioned by corruption and traditional politics. His ability to weave humour into serious discussions has remained a hallmark of his presidency.

Example: During the Russia-Ukraine conflict, Zelenskyy's speeches often included humour to boost morale. When addressing international aid delays, he joked, "Send me the planes. I'll even take them without pilots if you're busy." This blend of levity and urgency highlighted his determination while energizing his supporters.

Humour in Diplomacy: Breaking Barriers

Humour is not confined to domestic politics; it plays a significant role in international relations. A well-placed joke can ease tensions, foster goodwill, and create a foundation for productive negotiations.

Example: Ronald Reagan's Cold War Quip

During the Cold War, President Ronald Reagan famously joked, "We begin bombing in five minutes," while testing a microphone. Though the remark sparked controversy, Reagan's humour often humanized him, softening the image of the U.S. during tense global standoffs.

Example: Jacinda Ardern's Empathy and Humour
New Zealand Prime Minister Jacinda Ardern has used humour to connect with her citizens while navigating crises. After a viral video showed her struggling to pronounce a Māori name, Ardern joked, "I blame it on the baby brain. I've got an excuse now!" Her self-deprecating humour reinforced her relatability, making her leadership style both empathetic and effective.

The Risks of Humour in Leadership

While humour is a powerful tool, it is not without risks. Misjudged jokes or poorly timed quips can backfire, alienating audiences and undermining credibility.

Example: Boris Johnson's Controversial Comments
Boris Johnson, known for his flamboyant humour, has faced criticism for off-colour remarks. His description of women wearing burqas as "letterboxes" sparked outrage, illustrating how humour that targets vulnerable groups can erode trust and alienate constituents.

The key to effective humour in leadership lies in understanding the audience and context. A leader's humour must be inclusive, authentic, and aligned with their message to avoid unintended harm.

The Laughing Leader

From Lincoln's self-deprecating charm to Zelenskyy's modern-day quips, humour remains a timeless tool for leaders. It disarms, connects, and inspires, turning moments of tension into opportunities for unity. Leaders who wield humour wisely leave an indelible mark, transcending their policies to forge emotional bonds with their followers.

In the words of Theodore Roosevelt, "It is a fine thing to be cheerful. It is better to be humorous. It is best to be both." For leaders, the ability to laugh with—and not at—their audience is more than a skill; it is a legacy.

CHAPTER 10: LAUGHING THROUGH TEARS

Humour In Tragedy

In the shadow of calamity, humour often seems out of place. Yet, history and human behaviour reveal a paradoxical truth: laughter is as vital in moments of tragedy as it is in times of joy. From the trenches of World War I to the shared challenges of the COVID-19 pandemic, humour has offered solace, resilience, and connection when despair seemed insurmountable.

This chapter explores the remarkable role of humour during crises, examining how it helped people cope with wars, pandemics, and natural disasters. With real-life examples and psychological insights, we'll uncover why laughter persists even in the darkest moments—and how it lights the way to recovery.

The Psychological Power of Humour in Crisis

Humour serves as a psychological defence mechanism,

helping people confront fear, grief, and uncertainty. Sigmund Freud described humour as a way to transform anxiety into amusement, a coping strategy that allows the mind to process pain without succumbing to it.

Why We Laugh During Tragedy
1. Emotional Release: Laughter provides a safe outlet for pent-up emotions, easing the psychological burden of stress and grief.
2. Reframing the Situation: Humour helps people reinterpret dire circumstances, turning fear into something manageable.
3. Fostering Connection: Shared laughter creates solidarity, reminding individuals that they are not alone in their struggles.

Psychologists call this phenomenon "tragic optimism"—the ability to find meaning and even joy in the face of suffering. Humour is a key component of this mindset, offering a way to navigate life's harshest realities without losing hope.

Humour in the Trenches: World War I

In the harrowing trenches of World War I, soldiers faced unimaginable conditions: mud, cold, constant shellfire, and the ever-present spectre of death. Amid this chaos, humour emerged as a lifeline, providing soldiers with moments of levity and connection.

Trench Humour: Coping with the Unthinkable
Soldiers often used gallows humour—a form of dark comedy that confronts death and despair with wit. Trench newspapers, created by soldiers themselves, were

filled with satirical cartoons, jokes, and absurd stories. These publications, such as The Wipers Times, mocked the absurdity of war while offering a brief escape from its horrors.

Example: In one issue of The Wipers Times, a satirical advertisement read: "If you're tired of being shot at, why not try our new Invisible Cloak? Guaranteed 100% effective, except on Thursdays." The humour was both a critique of military bureaucracy and a way to lighten the psychological toll of constant danger.

Nicknames and Wordplay
Soldiers also gave humorous nicknames to their hardships. The relentless German artillery bombardment became known as "Jack Johnsons," a nod to the heavyweight boxer's powerful punches. These nicknames transformed terrifying experiences into something more bearable, demonstrating humour's ability to reframe fear.

The Blitz Spirit: Humour in World War II

During World War II, humour was not confined to the battlefield. Civilians in bombed-out cities like London used humour to endure the Blitz—a relentless campaign of German air raids.

Churchill's Rallying Wit
Winston Churchill's humour became a beacon of resilience for the British people. After a particularly devastating raid, Churchill quipped, "If Hitler invaded hell, I would make at least a favourable reference to the devil in the House of Commons." His sharp wit turned fear into defiance, inspiring a nation to hold its ground.

Daily Jokes in Bomb Shelters

Londoners waiting out air raids in underground shelters often exchanged jokes to pass the time. One popular quip went: "We keep moving because the Luftwaffe can't hit a moving target—unfortunately, that includes our wages." These jokes reflected the public's ability to find humour in their shared struggles, fostering a sense of community even in the darkest moments.

Humour in Pandemics: COVID-19 and Beyond

The COVID-19 pandemic brought unprecedented isolation and uncertainty, yet humour thrived as a coping mechanism worldwide. Social media became a hub for pandemic-related jokes, memes, and videos, creating a global community united by shared laughter.

Memes: The New Language of Resilience

From jokes about Zoom fatigue to the viral "quarantine haircut" memes, humour became a way for people to express frustration and find connection during lockdowns. One popular meme featured a picture of a man holding a sign that read, "Day 14: My housemates are trying to vote me off the island."

Memes about toilet paper shortages and sourdough baking turned collective anxieties into shared amusement. While the humour was light-hearted, it carried a deeper message: We're all in this together.

Dark Humour and Medical Professionals

Healthcare workers on the front lines of the pandemic often relied on dark humour to cope with the emotional toll of their work. A nurse shared a meme of a skeleton wearing a mask, captioned: "Me after trying to explain to one more person why they need to wear a mask." This humour, though tinged with frustration, provided a way to process stress and exhaustion.

Humour as Public Health Messaging
Governments and organizations also used humour to promote public health measures. In New Zealand, Prime Minister Jacinda Ardern's playful approach included jokes about the country's "team of five million" fighting the virus together. These light-hearted messages made serious topics more accessible, encouraging compliance without inducing panic.

Natural Disasters: Finding Light in the Darkness

Humour has also played a crucial role in helping communities recover from natural disasters. From hurricanes to earthquakes, laughter becomes a way to rebuild not just infrastructure but also collective morale.

Example: New Orleans After Hurricane Katrina
In the wake of Hurricane Katrina, New Orleans residents turned to their unique brand of humour to process the devastation. Mardi Gras floats poked fun at government mismanagement, with one float featuring a FEMA representative holding a sign that read, "We'll Be There in a Minute." This satire, while critical, allowed the community to channel their frustration into creativity, reaffirming their resilience.

Example: Earthquake Humour in Japan
After the 2011 Tōhoku earthquake and tsunami, Japanese citizens shared light-hearted anecdotes to lift spirits.

One joke involved a man whose house tilted during the quake, remarking, "Well, at least I don't have to level my floors anymore." These moments of humour provided emotional relief, helping survivors navigate the aftermath.

The Science of Humour in Crisis

Research confirms that humour has tangible psychological and physiological benefits during crises. Studies show that laughter reduces stress hormones like cortisol while increasing endorphins, which promote feelings of well-being. This chemical response makes humour a natural antidote to fear and anxiety.

Study: Humour in World War I
A 2015 study analysed letters and journals from World War I soldiers, finding that humour was a consistent coping mechanism across ranks. Soldiers who reported using humour also exhibited higher levels of emotional resilience, suggesting that laughter played a crucial role in maintaining mental health.

Study: Humour During COVID-19
During the COVID-19 pandemic, researchers at the University of Colorado found that people who engaged with humorous content reported lower levels of anxiety and loneliness. The study concluded that humour fosters connection and provides a sense of control, even in the face of uncertainty.

Why Humour Matters in Recovery

Humour not only helps people survive crises but also aids

in the recovery process. By reframing trauma, humour allows individuals and communities to move forward without being defined by their pain.

Example: The 9/11 Jokes That Came Later
In the years following the September 11 attacks, comedians like Jon Stewart and Tina Fey gradually reintroduced humour about the event. Their jokes, while respectful, allowed audiences to process their grief and begin the journey toward healing. Humour didn't trivialize the tragedy—it provided a way to confront it with courage and humanity.

Laughing Through the Storm

Humour in times of tragedy is not an escape; it is an act of defiance. It says, "We are still here. We are still human." Whether in the trenches of World War I, the bomb shelters of London, or the quarantined homes of a global pandemic, laughter has been a lifeline, offering hope, connection, and strength.

As we face future crises, we carry with us the lessons of humour's resilience. In laughter, we find the courage to endure and the reminder that even in our darkest moments, joy is never far away. In the words of Charlie Chaplin, "To truly laugh, you must be able to take your pain and play with it." And in that playfulness, humanity finds its greatest strength.

CHAPTER 11: THE ROLE OF SATIRE IN SOCIAL CHANGE

Satire is humour with purpose—a sharp-edged tool wielded to critique authority, expose hypocrisy, and challenge societal norms. By cloaking its critique in wit and absurdity, satire invites reflection and sparks dialogue, often succeeding where more straightforward approaches might fail. From Jonathan Swift's 18th-century essays to modern satirical platforms like The Onion, satire has remained a powerful vehicle for social change, adapting to new eras and technologies while retaining its core mission: to provoke thought through laughter.

This chapter delves into satire's enduring impact, tracing its history and examining its influence on culture, politics, and social movements.

What Is Satire?

Satire is humour with a mission. It uses exaggeration, irony, and absurdity to expose flaws in individuals, institutions, and ideologies. By holding a distorted mirror to society, satire forces its audience to confront uncomfortable truths in ways that are both engaging and disarming.

Why Satire Works
- Elicits Laughter and Thought: The humour draws people in, while the underlying critique keeps them engaged.
- Circumvents Resistance: By presenting criticism as entertainment, satire avoids the defensiveness that often accompanies direct confrontation.
- Amplifies Voices: Satire amplifies marginalized perspectives, giving voice to those who might otherwise go unheard.

Jonathan Swift: The Father of Satire

Jonathan Swift, the Irish author and clergyman, is often regarded as one of history's most influential satirists. His 1729 essay, A Modest Proposal, remains a quintessential example of how satire can highlight societal injustices.

Example: A Modest Proposal
In this essay, Swift suggests an outrageous solution to Ireland's famine and poverty: the poor should sell their children as food to the wealthy. The grotesque premise shocks the reader, but Swift's underlying critique of British exploitation and indifference toward the Irish is unmistakable. By couching his argument in dark humour, Swift forced his audience to grapple with the moral failures of their society.

Impact
While A Modest Proposal did not lead to immediate policy changes, it remains a cornerstone of satirical literature,

demonstrating satire's ability to provoke outrage and demand reflection. Its enduring relevance underscores satire's power to address systemic issues with clarity and wit.

Modern Satire: The Onion and The Daily Show

In the digital age, satire has found new homes on television and the internet, where platforms like The Onion and The Daily Show use humour to dissect current events and challenge power structures.

Example: The Onion
Known for its biting headlines, The Onion has become a staple of modern satire. During the 2008 financial crisis, it published a headline reading, "Recession-Plagued Nation Demands New Bubble to Invest In." This single sentence encapsulated the absurdity of the economic collapse, mocking both the financial industry and public complacency.

Impact: While The Onion rarely engages in direct activism, its satirical lens brings attention to issues often overlooked in traditional media. By distilling complex topics into humorous soundbites, it makes critique accessible to a wider audience.

Example: The Daily Show
Under Jon Stewart's leadership, The Daily Show became a cultural juggernaut, blending news and comedy to expose political hypocrisy. In 2004, Stewart famously appeared on CNN's Crossfire, a program that thrived on partisan bickering, and delivered a scathing critique of its format. "Stop hurting America," he implored the hosts, using

humour to call out the show's role in eroding meaningful discourse.

Impact: Stewart's appearance led to Crossfire's cancellation, showcasing satire's ability to influence media and reshape public conversations.

Satire and Social Movements

Satire has often been a rallying force for social change, amplifying the voices of marginalized communities and challenging systemic oppression.

Example: Charlie Hebdo and Free Speech
The French satirical magazine Charlie Hebdo has been both celebrated and criticized for its provocative humour, often targeting religion and politics. In 2015, the magazine became a global symbol of free speech after a terrorist attack on its offices. While some viewed its humour as offensive, others defended its right to challenge authority through satire.

Example: Black Lives Matter and Satirical Memes
During the Black Lives Matter protests, satirical memes became a digital tool for critique. One widely shared meme featured a mock advertisement for "Karen Repellent," poking fun at the phenomenon of individuals weaponizing their privilege to police others. These memes, while humorous, carried sharp commentary about systemic racism and privilege, galvanizing online discussions.

Comedians as Social Critics

In addition to traditional media, stand-up comedians have emerged as powerful satirists, using humour to tackle pressing social issues.

Example: Hannah Gadsby's Nanette

Hannah Gadsby's Netflix special Nanette blurred the lines between comedy and critique, addressing topics like gender, trauma, and systemic inequality. Gadsby used humour to draw in her audience before delivering raw, unflinching truths, challenging the conventions of stand-up comedy itself.

Impact: Nanette sparked global conversations about representation and the role of comedy in addressing societal pain, proving that satire can be deeply personal and transformative.

Example: Hasan Minhaj's Patriot Act

Hasan Minhaj's Patriot Act combined humour with investigative journalism to explore topics like student debt, immigration, and authoritarianism. Minhaj's ability to balance comedy with in-depth analysis made the show a unique platform for educating audiences while entertaining them.

Impact: Minhaj's work has been praised for its accessibility, using humour to engage younger audiences with complex political issues.

The Risks and Challenges of Satire

While satire can be a force for good, it is not without risks. Poorly executed satire can alienate audiences, perpetuate stereotypes, or even incite harm.

Example: The Charlie Hebdo Controversy

Critics argue that some of Charlie Hebdo's cartoons crossed the line into cultural insensitivity, reinforcing stereotypes rather than challenging power. This raises an important question: Where is the line between satire and offense?

Satire in a Polarized World

In today's highly polarized climate, satire often preaches to the choir, reinforcing existing beliefs rather than bridging divides. Satirists must navigate this challenge carefully, ensuring their work fosters dialogue rather than deepening divisions.

The Future of Satire

As technology evolves, so too does satire. The rise of AI-generated humour, deepfake videos, and virtual reality presents new opportunities and challenges for satirists. However, the essence of satire—its ability to critique and provoke—remains timeless.

Laughing Toward Change

Satire is more than entertainment; it is a call to action. By exposing the absurdities of power and the hypocrisies of society, satire holds up a mirror to humanity, challenging us to reflect and improve. From Jonathan Swift's biting essays to Hasan Minhaj's modern-day storytelling, satire has consistently proven its ability to spark change through laughter.

In the words of Oscar Wilde, "If you want to tell people

the truth, make them laugh—otherwise, they'll kill you." Satire's enduring power lies in its ability to do just that: to reveal the truth, one laugh at a time.

CHAPTER 12: HUMOUR IN MEDICINE

A Laugh A Day Keeps The Doctor Away

The old saying, "Laughter is the best medicine," may sound like a cliché, but scientific research and real-life experiences continue to affirm its truth. Humour is more than a pleasant distraction in healthcare—it's a potent tool that improves patient outcomes, enhances recovery, and strengthens the doctor-patient relationship. In medical settings, where stress and fear often prevail, humour can transform the atmosphere, offering hope and healing in ways that medication alone cannot.

This chapter explores the profound role of humour in medicine, examining its effects on physical and mental health, its applications in patient care, and the studies that underscore its therapeutic potential.

The Science of Laughter in Medicine

Laughter triggers a cascade of physiological and

psychological effects, making it a natural complement to traditional medical treatments. When we laugh, the body releases endorphins—natural painkillers that elevate mood and promote well-being. At the same time, laughter reduces levels of cortisol, the stress hormone, alleviating anxiety and tension.

Laughter's Physical Benefits
1. Boosts Immunity: Laughter enhances the production of antibodies and activates T-cells, improving the body's ability to fight infections.
2. Reduces Pain: By releasing endorphins, laughter increases pain tolerance, providing relief to patients with chronic conditions.
3. Improves Circulation: Laughing increases heart rate and oxygen flow, benefiting cardiovascular health.
4. Enhances Relaxation: After a bout of laughter, the body experiences muscle relaxation, reducing physical tension.

Laughter's Psychological Benefits
- Promotes emotional resilience in the face of illness.
- Fosters a sense of connection and reduces feelings of isolation.
- Encourages positive thinking, which has been linked to faster recovery times.

Humour in Patient Care

In healthcare settings, humour serves as a bridge between patients and providers, breaking down barriers and creating trust. For patients, humour can alleviate the fear and discomfort associated with medical procedures. For providers, humour offers a way to connect with patients on a human level, reminding both parties that healing is not just a clinical process—it's a deeply personal one.

Example: The Clown Doctors
In hospitals around the world, professional medical clowns, or "clown doctors," visit paediatric wards to engage young patients in playful activities. These interactions, which often involve jokes, magic tricks, and silliness, have been shown to reduce anxiety and improve recovery outcomes for children undergoing surgery or long-term treatment.

Study: A 2011 study published in Paediatrics found that children who interacted with clown doctors before surgery experienced lower levels of preoperative anxiety compared to those who did not. The laughter provided a distraction, shifting the focus from fear to joy.

Example: A Nurse's Joke
During a routine blood draw, a nurse noticed her patient fidgeting nervously. "Don't worry," she said with a grin, "I've been practicing on oranges all morning!" The patient laughed, and the tension in the room dissolved. This simple moment of humour not only eased the patient's fear but also created a sense of trust.

Humour and Chronic Illness

For patients living with chronic illnesses, humour provides a way to reclaim agency and cope with the emotional toll of long-term treatment. Laughter doesn't cure disease, but it makes the journey more bearable, fostering resilience and improving quality of life.

Example: Norman Cousins
Norman Cousins, a journalist diagnosed with a debilitating autoimmune disease in the 1960s, famously credited laughter with aiding his recovery. Dissatisfied with traditional treatments, Cousins watched hours of comedy films, including Marx Brothers classics, and reported that ten minutes of belly laughter provided two hours of pain-free sleep. His remarkable improvement, chronicled in Anatomy of an Illness, inspired further research into the therapeutic power of humour.

Impact: Cousins' case became a landmark example of how humour can complement medical treatment, sparking the rise of laughter therapy as a legitimate field of study.

Studies on Laughter Therapy

Research continues to affirm the benefits of humour in medicine. Laughter therapy, which involves deliberate activities to induce laughter, has been integrated into programs for patients with cancer, heart disease, and mental health conditions.

Study: Cancer and Laughter
A study published in Supportive Care in Cancer found that cancer patients who participated in laughter therapy sessions reported reduced levels of pain, fatigue, and emotional distress. The therapy sessions, which included comedic skits and laughter exercises, also improved patients' overall mood and sense of well-being.

Study: Stress Reduction in Caregivers
Caregivers, who often experience high levels of burnout, have also benefited from laughter therapy. A 2016

study in The Journal of Nursing Education and Practice found that caregivers who engaged in regular laughter exercises reported lower stress levels and improved job satisfaction.

Real-Life Stories: Humour in Healing

A Comedian in the Chemo Ward
When stand-up comedian Tig Notaro was diagnosed with breast cancer, she turned her experience into a comedy routine, joking about her diagnosis and the absurdities of her treatment. Her performance, titled Live, became a viral sensation, resonating with audiences worldwide. Notaro's humour not only helped her cope but also inspired others facing similar challenges, proving that laughter can coexist with even the most serious struggles.

A Doctor's Prescription for Laughter
Dr Patch Adams, a physician and social activist, famously incorporated humour into his medical practice, dressing as a clown and using jokes to connect with patients. His philosophy, popularized by the 1998 film Patch Adams, emphasized the importance of treating the whole person, not just the disease. While unconventional, his approach highlighted the therapeutic potential of humour in humanizing medicine.

Humour in End-of-Life Care

Even at the end of life, humour remains a vital part of the healing process. Hospice care providers often use humour to bring comfort to patients and their families, creating moments of joy amid grief.

Example: A Light-hearted Goodbye
A hospice patient nearing the end of life joked with her nurse, "When I get to heaven, I'll put in a good word for you—but only if they serve chocolate cake!" The nurse laughed, and the patient's family joined in, transforming a sombre moment into a celebration of life. This exchange illustrates how humour can ease emotional pain, fostering connection and closure.

Humour's Place in Modern Medicine

As healthcare continues to evolve, humour is being recognized as a critical component of holistic care. From laughter yoga classes to humour therapy apps, the integration of humour into medical practice is gaining momentum.

Humour in Telemedicine
In the age of telemedicine, humour has adapted to virtual interactions. Doctors and patients often exchange light-hearted banter during video calls, breaking the monotony of remote consultations. One doctor quipped to a patient, "At least this time you don't have to worry about bad hospital food!"

Humour and AI in Healthcare
As artificial intelligence becomes a larger part of healthcare, developers are exploring ways to incorporate humour into AI systems. Chatbots programmed with light-hearted responses can ease patient anxiety, creating a friendlier interface for medical queries.

The Risks of Humour in Medicine

While humour has immense benefits, it must be used thoughtfully. Poorly timed or inappropriate jokes can undermine trust and cause discomfort. For example, a joke that trivializes a patient's condition may come across as dismissive rather than supportive. Medical professionals must strike a balance, ensuring their humour is sensitive to the context and the individual.

A Prescription for Laughter

Humour in medicine is not a luxury—it is a necessity. It soothes pain, strengthens relationships, and uplifts spirits, transforming healthcare into a more human and compassionate experience. As patients laugh their way through fear and uncertainty, they find strength, hope, and connection in the process.

In the words of Norman Cousins, "Laughter is a form of internal jogging. It moves your internal organs, enhances respiration, and invigorates the mind." Whether through a comedian in a chemo ward, a doctor's light-hearted quip, or a child laughing with a clown doctor, humour proves time and again that healing is more than a physical act—it is a celebration of life itself.

CHAPTER 13: CULTURAL LAUGHTER

Humour Across The Globe

Humour is universal, but the ways we express and interpret it are shaped by our cultures, histories, and traditions. Across the globe, humour manifests in unique styles, from the refined wordplay of Japan to the playful sarcasm of Britain and the storytelling wit of India. These variations reflect not just differences in comedic taste but also the deeper values, norms, and struggles of each society.

In this chapter, we'll explore humour's diverse forms, comparing styles from around the world and illustrating how laughter transcends boundaries while celebrating cultural distinctiveness.

The Universality of Humour

Though its expressions differ, humour serves similar purposes across cultures. It fosters social bonds, diffuses tension, and provides relief from life's challenges. What varies is the lens through which humour is understood. Cultural contexts influence what people find funny, from the types of jokes they enjoy to how and when they express laughter.

The Framework of Cultural Humour
1. Wordplay and Linguistics: Languages with intricate structures often inspire humour through puns and double meanings.
2. Social Commentary: Humour reflects societal norms, allowing communities to critique themselves and others.
3. Physical Comedy: Universally understood, physical humour transcends language barriers but adapts to local traditions.

Japanese Humour: The Subtlety of Wordplay

Japanese humour often emphasizes linguistic creativity and situational irony, favouring clever wordplay over overt physical comedy. This style reflects Japan's cultural values of harmony and subtlety, where humour often avoids direct confrontation or ridicule.

Example: The Art of "Oyaji Gags"
"Oyaji gags" are dad jokes with a Japanese twist, relying heavily on puns. For instance:
- "I'm kanpai (cheers), but not kanpai (dry)!"
This pun juxtaposes similar-sounding words with different meanings, eliciting groans and laughter.

Manzai Comedy
Manzai, a traditional Japanese comedic style, features two performers: a boke (the fool) and a tsukkomi (the straight

man). The boke makes absurd statements or mistakes, and the tsukkomi reacts with exaggerated corrections. This dynamic reflects Japan's appreciation for balance and structured chaos.

Cultural Insight: Japanese humour thrives on subtlety and shared understanding, reinforcing social bonds without disrupting harmony.

Indian Humour: Storytelling and Satire

Indian humour is deeply rooted in storytelling traditions, where jokes and anecdotes are often infused with wit, wordplay, and social commentary. This reflects India's rich oral history and its ability to find humour even in adversity.

Example: Political Satire
Indian comedians often tackle politics with biting humour. For instance, during election seasons, memes and stand-up routines mock campaign promises, highlighting contradictions with sharp wit. One comedian joked, "Indian politicians love bridges. They promise one even if there's no river!"

Bollywood and Humour
Bollywood films often combine slapstick, wordplay, and situational comedy to create moments of levity. Movies like Hera Pheri and 3 Idiots mix humour with poignant messages, showcasing the Indian knack for blending laughter with life lessons.

Cultural Insight: Indian humour is both inclusive and layered, reflecting a society that embraces diversity and uses laughter to navigate complexities.

British Humour: Sarcasm and Self-Deprecation

British humour is famous for its dry wit, sarcasm, and self-deprecation. Rooted in centuries of literary and theatrical tradition, it often pokes fun at societal norms, authority, and the self.

Example: Self-Deprecation
A quintessential British trait is the ability to laugh at oneself. When London hosted the 2012 Olympics, the opening ceremony featured James Bond escorting the Queen in a parachute stunt—a humorous nod to British stoicism and eccentricity.

The Role of Satire
From Shakespeare's plays to Monty Python's Flying Circus, British humour has consistently used satire to critique power and convention. One iconic Monty Python sketch features a man arguing with a clerk about whether he paid for a conversation, a sharp commentary on bureaucracy's absurdity.

Cultural Insight: British humour values subtlety and intelligence, favouring wit over overt displays of emotion.

American Humour: Bold and Relatable

American humour tends to be direct and larger-than-life, reflecting a culture that values individuality and relatability. Comedy in the United States often combines observational humour with bold physicality, appealing to broad audiences.

Example: Stand-Up Comedy

American stand-up comedians like Jerry Seinfeld and Kevin Hart excel in observational humour, drawing laughs from the quirks of everyday life. Seinfeld's iconic bit about airplane peanuts—"They give you the smallest snack for the longest flight!"—highlights humour's ability to make the mundane hilarious.

Late-Night Satire

American late-night shows like Saturday Night Live and The Daily Show blend humour with political critique, shaping public opinion while entertaining audiences. During the 2020 election, jokes about mail-in voting and political debates became viral, offering comedic relief amid tension.

Cultural Insight: American humour is bold, accessible, and deeply tied to current events, reflecting a society that values expression and relatability.

Middle Eastern Humour: Resilience Through Laughter

In the Middle East, humour often serves as a means of coping with conflict and hardship. Satirical sketches, stand-up routines, and folk humour address sensitive topics with warmth and wit.

Example: Satire in Syria

During Syria's civil war, comedians used humour to critique both local and international responses. A viral skit featured actors pretending to "fix" the country's infrastructure by painting over cracks, a biting metaphor for superficial solutions.

Folk Tales and Jokes
Traditional Middle Eastern humour often revolves around the character of Juha, a wise fool whose misadventures carry moral lessons. In one tale, Juha tells people he can make his donkey sing. When they pay to see it, he says, "You didn't ask when it would sing—maybe tomorrow, maybe never!"

Cultural Insight: Humour in the Middle East balances levity with resilience, offering hope and critique in equal measure.

African Humour: Community and Spirit

African humour is rich in storytelling and communal laughter, reflecting the continent's diverse cultures and emphasis on shared experience.

Example: Nigerian Comedy
In Nigeria, stand-up comedians like Basketmouth and Ali Baba use humour to address societal issues such as corruption and generational conflicts. Jokes about traffic jams and everyday struggles resonate widely, turning shared frustrations into laughter.

Folktales and Tricksters
Traditional African folktales often feature trickster characters like Anansi the Spider, who outsmarts others through cleverness. These stories, passed down orally, use humour to teach lessons about ingenuity and resilience.

Cultural Insight: African humour thrives on communal participation, turning shared challenges into opportunities for connection.

Humour That Transcends Boundaries

Despite cultural differences, certain types of humour —such as physical comedy and universal absurdities— transcend borders. A silent Charlie Chaplin routine can evoke laughter in Tokyo as easily as in New York, demonstrating humour's ability to unite.

Example: Viral Memes
In the digital age, memes have become a global language of humour. A meme about Zoom call fails during the COVID-19 pandemic resonated universally, as people worldwide laughed at the shared absurdities of remote work.

The Role of Humour in Bridging Cultures

Humour is not just a reflection of culture—it is also a bridge between them. When people laugh together, they transcend differences, finding common ground in shared amusement.

Example: Cross-Cultural Comedians
Comedians like Trevor Noah, whose humour draws on his South African heritage, have bridged cultural gaps by highlighting universal themes. Noah's observations about cultural misunderstandings—such as how different nations interpret punctuality—elicit laughter while fostering cross-cultural understanding.

Laughter Without Borders

Humour is both deeply local and universally human. It reflects the values, struggles, and quirks of individual cultures while reminding us of our shared humanity. Whether through Japanese puns, Indian wit, British sarcasm, or American boldness, humour connects us across languages and borders, proving that laughter truly knows no limits.

In the words of Charlie Chaplin, "Laughter is the universal language." As we laugh together—at jokes, stories, or memes—we celebrate not just our differences but also the common threads that bind us all.

CHAPTER 14: HUMOUR AND IDENTITY

What Our Laughs Say About Us

Humour is not just a moment of amusement; it is a profound expression of who we are. The things that make us laugh reflect our identities—our values, struggles, and the social groups to which we belong. From ethnic humour that navigates cultural pride and pain to generational jokes that encapsulate the spirit of an era, humour serves as both a mirror and a bridge. It can unite communities, foster self-awareness, and affirm individuality, all while eliciting laughter.

In this expanded exploration of humour and identity, we will delve deeper into how our laughs reveal the intricacies of personal and collective identities, examining real-life examples of ethnic, generational, and social humour.

Humour as a Reflection of Identity

Laughter is a universal human experience, but what we find funny is deeply personal and shaped by the social contexts in which we live. Humour reflects both our individual personalities and the cultural frameworks that influence our worldview.

Why Humour Reveals Identity
1. Personal Identity: The jokes we tell or enjoy highlight aspects of our character, including whether we are more drawn to sarcasm, slapstick, or intellectual humour.
2. Cultural Identity: Humour specific to a cultural or ethnic group often explores shared experiences, historical struggles, and societal quirks.
3. Group Identity: Inside jokes and communal humour reinforce bonds within specific communities, acting as a badge of belonging.

When we laugh, we are signalling what resonates with us

—and by extension, revealing parts of who we are.

Ethnic Humour: Pride, Pain, and Connection

Ethnic humour occupies a unique space in comedy. When created and appreciated within a community, it becomes a celebration of shared experiences and a tool for coping with historical challenges. When misused or directed at a group from the outside, it risks reinforcing stereotypes and alienating its audience.

Celebrating Culture Through Humour
Ethnic humour often draws on cultural traditions, shared struggles, and unique quirks, turning them into moments of collective joy and self-reflection.

- Example: Jewish Humour and Resilience
Jewish humour is renowned for its self-deprecating wit and philosophical undertones. It often blends existential musings with comedic absurdity, reflecting a history marked by resilience in the face of adversity. For instance, the classic joke, "Why do Jewish divorces cost so much? Because they're worth it!" playfully critiques both marital dynamics and cultural values around family.

- Example: Indian-American Comedy and Bicultural Identity
Comedians like Hasan Minhaj and Mindy Kaling weave their bicultural experiences into their humour, navigating the push-and-pull of immigrant identity. Minhaj's quip about his father's thriftiness—"He would bargain at Costco!"—resonates with audiences familiar with the frugal habits of first-generation parents, turning a shared cultural trait into a source of laughter and

connection.

The Line Between Humour and Harm

Ethnic humour's power to affirm identity also makes it a delicate tool. Misjudged or externally imposed ethnic jokes can perpetuate stereotypes, alienating rather than affirming.

- Example: Misplaced Ethnic Humour
In a professional setting, a non-Hispanic coworker attempts to make a joke about a colleague's "spicy food preference," intending to bond but instead invoking a tired stereotype. Such humour, when detached from context or mutual understanding, risks reinforcing negative tropes rather than fostering camaraderie.

Takeaway: Ethnic humour is most powerful when it comes from within the community, offering both critique and celebration.

Generational Humour: Defining an Era

Each generation develops its own humour style, shaped by cultural milestones, technological advancements, and societal shifts. From Baby Boomers' love of situational comedy to Gen Z's embrace of absurdism, generational humour offers insight into the priorities and anxieties of the era.

Boomer Humour: Traditional and Relatable

Baby Boomers grew up with humour rooted in sitcoms like I Love Lucy and All in the Family, where exaggerated characters and situational gags provided family-friendly laughs. This humour reflects the optimism and collective values of the post-war period, with a focus on relatable everyday situations.

Gen X Humour: Irony and Sarcasm

Gen X, raised during a time of shifting societal norms, developed a more ironic sense of humour. Shows like The Simpsons and Seinfeld epitomized this era's humour, poking fun at authority, conventionality, and even the idea of humour itself. A Gen X joke might quip, "It's not that I don't care—I just care ironically."

Millennial Humour: Relatable Struggles

Millennials, coming of age during economic instability and rapid technological change, use humour to process shared anxieties. Memes like "adulting is hard" reflect the generation's struggle with student debt, workplace challenges, and delayed milestones. One popular meme jokes, "I can't afford therapy, so I just search 'how to handle crushing existential dread' on Google."

Gen Z Humour: Absurd and Existential

Gen Z humour thrives on surrealism and chaos, often embracing randomness as a comedic device. TikTok videos and viral memes like "Shrek as a life coach" exemplify the generation's ability to laugh at the absurdity of modern life. This reflects a worldview shaped by uncertainty, where humour becomes both a critique and a coping mechanism.

Inside Jokes: The Language of Belonging

Inside jokes are a form of humour that operates within a closed group, creating a sense of exclusivity and intimacy. These jokes, often incomprehensible to outsiders, strengthen bonds by affirming shared experiences.

Example: Family Humour
In one family, a toddler's mispronunciation of "ambulance" as "ambliance" became an enduring inside joke. Years later, whenever someone needed first aid, another family member would say, "Call the ambliance!" eliciting laughter from those in the know.

Example: Workplace Humour
In an office with a frequently malfunctioning coffee machine, employees began calling it "The Bean Conspiracy." The nickname turned a shared annoyance into a running joke, fostering a sense of unity among the staff.

Takeaway: Inside jokes highlight the human need for connection, turning everyday moments into symbols of belonging.

Humour and Social Identity

Beyond ethnicity and generation, humour also reflects broader aspects of identity, such as gender, class, and sexuality. Comedians often use humour to challenge norms, critique inequalities, and celebrate diversity.

Example: Feminist Humour
Comedians like Ali Wong use humour to critique traditional gender roles. Wong's line, "I don't want to lean in—I want to lie down," turns a common feminist slogan into a playful rejection of burnout culture, resonating with audiences tired of societal expectations.

Example: LGBTQ+ Humour
Queer humour often blends satire and celebration,

reflecting resilience and creativity. Shows like RuPaul's Drag Race feature playful humour that critiques heteronormativity while celebrating LGBTQ+ identities.

Insight: Humour becomes a tool for empowerment, allowing marginalized groups to assert agency and reshape narratives.

The Risks and Rewards of Humour and Identity

While humour can affirm and connect, it can also divide or harm. Misjudged jokes, especially those that perpetuate stereotypes, risk alienating their audience.

Example: Gendered Humour Gone Wrong
At a corporate event, a male executive jokingly told a female colleague, "Don't overthink it; leave the math to us." Though he intended it as playful, the comment reinforced harmful gender stereotypes, creating discomfort and damaging trust.

Takeaway: Humour rooted in identity must navigate context, intent, and audience to avoid harm.

The Unifying Power of Humour

Despite its complexities, humour often transcends boundaries, creating shared moments of laughter that bridge divides. Viral memes, multicultural comedy shows, and cross-generational jokes demonstrate humour's ability to connect us.

Example: The IKEA Meme
A meme about the struggles of assembling IKEA

furniture—captioned "If you can build this, you can survive anything"—resonates globally. The humour is universal, highlighting shared human experiences.

Our Laughs, Our Selves

Humour reflects identity, revealing who we are as individuals and as members of larger communities. From ethnic jokes that navigate cultural pride and pain to generational memes that capture an era's spirit, our laughter tells a story. It affirms our values, strengthens our bonds, and reminds us of our shared humanity.

In the words of Maya Angelou, "I don't trust anyone who doesn't laugh." Through humour, we express ourselves, connect with others, and celebrate the beautiful complexity of life—all while sharing a good laugh.

CHAPTER 15: THE FUTURE OF LAUGHTER

Humour In A Changing World

As humanity stands on the cusp of profound transformations driven by artificial intelligence, globalization, and societal change, humour continues to adapt and evolve. Laughter, that universal expression of joy and connection, faces new challenges and opportunities in this brave new world. Can robots understand humour? Will globalization create a universal comedic language, or will it dilute cultural distinctiveness? And as society grapples with crises of identity, technology, and inequality, will humour remain a source of resilience and connection?

This chapter explores the trajectory of humour in an AI-driven, interconnected world, addressing its role in shaping the future of human relationships, technology, and cultural exchange.

Humour in an AI-Driven World: Can Robots Truly "Laugh"?

Artificial intelligence (AI) has already made strides in creating and curating humour, yet the deeper question persists: Can machines ever truly understand or feel humour? While AI can generate jokes, memes, and even laughter-like sounds, its capabilities are bound by algorithms and datasets—mimicking, rather than experiencing, humour.

AI-Generated Jokes: Clever, But Lacking Depth
AI systems like GPT-3 and Bard can produce jokes that follow familiar patterns:
- Why don't robots ever panic? They're programmed to stay calm under "pressure sensors."

While amusing, these jokes often lack the nuance and spontaneity of human humour. They are formulaic, reflecting patterns from datasets but missing the emotional and contextual layers that give humour its depth.

Humour Relies on Context and Empathy
Understanding humour requires grasping the subtleties of context, timing, and emotional cues. For instance, a joke about procrastination might resonate differently in a high-pressure work environment versus a casual social gathering. Machines, which lack lived experiences, struggle to navigate these nuances.

Robots That "Laugh"
In Japan, researchers at Hiroshi Ishiguro Laboratories created Erica, a humanoid robot programmed to simulate

laughter during conversations. Using cues such as tone and facial expressions, Erica produces an artificial chuckle when it detects humour. However, this laughter is purely performative, a calculated response devoid of genuine understanding or feeling.

The Philosophical Barrier: Can AI Understand Humour?

At its core, humour is a human phenomenon, deeply tied to consciousness, emotion, and the complexities of social interaction. While machines can mimic humour, they cannot replicate the lived experiences that give laughter its meaning. Humour often arises from contradictions, shared absurdities, and vulnerabilities—qualities that are intrinsically human.

Humour as a Measure of Humanity

If humour reflects the human condition, then its absence in AI underscores the gap between machine intelligence and human consciousness. A robot may generate jokes, but it cannot laugh with joy, relief, or camaraderie. In this way, humour becomes a defining marker of what it means to be human.

Globalization and the Rise of Universal Humour

In our interconnected world, humour is increasingly transcending cultural and linguistic barriers. Social media platforms like TikTok, Instagram, and Twitter have created a shared digital space where jokes, memes, and comedic videos can spread instantaneously. While this fosters a sense of global community, it also raises questions about the preservation of cultural distinctiveness in humour.

The Language of Global Humour: Memes

Memes have become the lingua franca of internet humour. Simple visuals paired with captions convey jokes that resonate universally, such as:

- The "Distracted Boyfriend" meme, where a man looking at another woman becomes a metaphor for countless scenarios, from procrastination to political critiques.
- The "Is This a Pigeon?" meme, where a character mistakenly identifies an object, highlights universal misunderstandings.

Memes thrive because they distil humour into its simplest form, relying on imagery and relatability rather than linguistic complexity.

The Tension Between Global and Local Humour

While global humour fosters connection, it risks flattening the diversity of comedic traditions. A Japanese pun, for example, may lose its meaning when translated, while British sarcasm might be misinterpreted in cultures that prioritize direct communication. Global humour must strike a balance between universality and specificity to avoid diluting cultural identities.

Cultural Exchange in Comedy

Globalization has also facilitated the exchange of comedic styles, introducing audiences to new perspectives. Netflix stand-up specials by comedians from Nigeria, India, and South Korea showcase the richness of international humour, proving that laughter can bridge cultural divides.

Humour as a Coping Mechanism in an Uncertain World

As humanity confronts challenges such as climate change, political instability, and social inequality, humour remains a critical tool for resilience. In times of uncertainty, laughter provides a sense of control, relief, and connection.

Dark Humour in Crisis
Dark humour, which finds comedy in adversity, has long been a coping mechanism during difficult times. For example:
- During the COVID-19 pandemic, memes about toilet paper shortages and Zoom mishaps went viral, offering levity amid global anxiety.
- A popular quip during climate change debates: "When the sea levels rise, at least we'll all get beachfront property!"

Dark humour allows individuals to confront fear and frustration with a sense of irony, turning despair into shared resilience.

Activism Through Humour
Humour is increasingly becoming a tool for activism, helping to challenge authority and engage audiences. Satirical shows like Last Week Tonight with John Oliver and online platforms like The Onion use humour to critique systemic injustices while informing and mobilizing viewers. Activist memes and satirical videos often spread faster than traditional news, reaching younger, tech-savvy audiences.

Speculation: The Future of Humour and Technology

The integration of AI, virtual reality (VR), and augmented

reality (AR) into our lives raises intriguing possibilities for the future of humour. How will these technologies shape the way we create and experience laughter?

AI-Personalized Comedy

Imagine AI systems that curate jokes, memes, and comedy routines tailored to individual preferences. These systems could analyse a user's humour style—whether they prefer dry wit, slapstick, or wordplay—and deliver content that resonates uniquely with them.

Humour in Virtual Reality

In virtual reality, comedy could become an immersive experience. Stand-up routines might allow audience members to interact with performers, or users could create personalized comedic scenarios within virtual worlds. For example, a VR program might simulate an exaggerated version of everyday frustrations—like a coffee machine that spouts philosophy instead of coffee—turning mundane annoyances into laughter-inducing experiences.

Humour as Therapy

AI and VR could also harness humour for therapeutic purposes. Laughter therapy programs might use AI-driven avatars to deliver jokes, engage in playful banter, or simulate comedic scenarios tailored to alleviate stress and anxiety.

Ethical Challenges

As humour becomes increasingly mediated by technology, ethical questions arise. Will AI humour reinforce stereotypes if trained on biased datasets? Who owns the intellectual property of AI-generated jokes? And how do we ensure that humour remains inclusive and respectful in an age of automation?

The Resilience of Human Humour

Despite technological advancements, humour remains one of the most human aspects of our experience. It is shaped by emotion, shared history, and the ability to find joy in life's contradictions. Machines may generate laughter-like responses, but they lack the spontaneity, empathy, and depth that define true humour.

The Essence of Human Laughter
Laughter is more than a reaction; it is a bond. It connects us to others, reflects our humanity, and reminds us of our resilience. As the world changes, humour will continue to evolve, adapting to new contexts while preserving its core essence: the ability to bring people together.

Laughing Into the Future

The future of humour lies at the intersection of tradition and innovation. As technology transforms the way we create and share comedy, laughter will remain a uniquely human experience, rooted in emotion and connection. Whether through AI-generated memes, virtual reality stand-up, or age-old jokes passed around a dinner table, humour will continue to adapt to the needs of a changing world.

Can robots ever truly laugh? Perhaps not in the way humans do. But as we explore that question, we reaffirm what makes humour so vital: its capacity to unite us, heal us, and remind us that even in the face of uncertainty, laughter endures.

In the words of Charlie Chaplin, "A day without laughter is a day wasted." As we step into an AI-driven, globalized future, we carry with us the timeless gift of humour, ensuring that no matter how the world changes, we will always have reasons to laugh.

CHAPTER 16: LAUGHING THROUGH TEARS

Humour In Adversity

Humour thrives where you'd least expect it. In the trenches of war, in the midst of pandemics, and in personal moments of heartbreak, laughter emerges like a stubborn flower cracking through concrete. Humour, in these moments, is not a dismissal of suffering but an act of defiance—a refusal to let pain have the final word.

This chapter explores how humour serves as a lifeline during crises, a tool that transforms despair into resilience. Through real-life stories from history and modern times, we'll see how laughter becomes a companion in adversity, offering relief, connection, and a glimmer of hope.

The Trenches of WWI: Gallows Humour in the Line of Fire

The muddy, perilous trenches of World War I were

places of unimaginable horror. Soldiers faced constant bombardment, the spectre of gas attacks, and the grim reality of losing comrades. Yet even in the shadow of death, humour found a way to survive.

Trench Humour: Laughing to Survive
Trench humour was often dark, irreverent, and deeply human. Soldiers mocked the absurdity of their circumstances, turning their dire situations into moments of shared levity. They nicknamed lice "chatting partners" and called their daily stew "trench soup," joking that its mysterious lumps must be rat tails.

Christmas Truce and Soccer
One of the most famous examples of humour in the trenches occurred during the Christmas Truce of 1914, when German and Allied soldiers emerged from their trenches to exchange gifts, share jokes, and even play a soccer match. The laughter they shared transcended language and politics, reminding them, however briefly, of their shared humanity.

Cartoons and Satirical Songs
Soldiers also turned to humour as a form of creative expression. Trenches were often adorned with caricatures of commanding officers or crude jokes about the enemy. Songs like Mademoiselle from Armentières, with its cheeky lyrics, became anthems of camaraderie, helping soldiers endure the chaos with a smile.

Humour in the Pandemic: Memes as Emotional Relief

In the face of COVID-19, humour once again emerged as a coping mechanism. As the world grappled with lockdowns, isolation, and uncertainty, people turned to the digital landscape to find solace in shared laughter.

Pandemic Memes: Humour in Isolation

Memes became a universal language during the pandemic, capturing shared experiences with wit and relatability. From jokes about Zoom meetings ("This meeting could have been an email") to the infamous sourdough bread craze, these digital quips offered a way to process the absurdity of the situation.

Toilet Paper Shortage Jokes

When panic buying led to toilet paper shortages, the internet responded with humour. Images of empty shelves captioned "Trading one roll for a three-bedroom house" went viral, transforming frustration into collective amusement. These jokes didn't solve the crisis, but they lightened the emotional burden, reminding people that they weren't alone.

Humour in Global Quarantine

In Italy, residents on lockdown stepped out onto their balconies to sing, play music, and tell jokes. Videos of these impromptu performances spread worldwide, showcasing humour as a way to foster connection and resilience even in isolation.

Personal Trauma: Finding Strength in Laughter

Humour isn't just a societal coping tool—it's deeply personal, helping individuals navigate grief, loss, and hardship. When life delivers its harshest blows, laughter can become a means of reclaiming agency and finding strength.

Comedians and Personal Pain

Many comedians draw on their own experiences of trauma to create humour that resonates deeply. Tig Notaro, a stand-up comedian, famously turned her cancer diagnosis into a groundbreaking comedy set. In her routine, she opened with the line, "Hello. I have cancer." The audience's initial shock turned into laughter as Notaro skilfully blended vulnerability with wit, showing that humour could coexist with profound pain.

"Comedy is tragedy plus time," said Carol Burnett, encapsulating the idea that humour allows us to revisit painful experiences with new perspective and resilience.

Laughter in Loss

In grief, humour often serves as a way to remember loved ones not just with sadness but with joy. At a funeral for her grandmother, a woman shared a story about how her grandma once "repaired" a broken lamp with duct tape and prayer. The room erupted in laughter, transforming a moment of sorrow into one of warmth and connection.

Humour as Resistance: Laughing in Oppression

Humour has long been a tool of resistance for oppressed communities, offering a way to push back against injustice and reclaim dignity.

Jewish Humour During the Holocaust

Even in the darkest hours of the Holocaust, humour persisted among Jewish communities as a form of spiritual resistance. In ghettos and concentration camps, people told jokes about their captors, using humour to assert their humanity in the face of dehumanization.

A Ghetto Joke

One joke told in the Warsaw Ghetto went: "Why are all our streets so clean? Because every week, the Gestapo sweeps them." The dark humour didn't erase the terror, but it allowed people to process their fear through laughter, giving them a momentary sense of control.

Civil Rights Movements

Humour also played a crucial role during the civil rights movement in the United States. Activists like Dick Gregory used stand-up comedy to highlight racial injustice, blending sharp humour with hard truths. Gregory's humour not only entertained but educated, challenging audiences to confront uncomfortable realities with a smile.

The Psychology of Humour in Adversity

Why does humour thrive in hardship? Psychologists suggest that laughter serves as a release valve for emotional tension. By reframing challenges through a comedic lens, humour helps individuals and groups regain a sense of agency and perspective.

Cognitive Reframing Through Humour

Humour allows people to reinterpret painful situations, finding absurdity in the unthinkable. This reframing helps reduce the psychological burden of trauma, offering a way to process emotions without becoming overwhelmed.

Study Insight: A 2016 study published in Emotion found that people who used humour to cope with stress experienced lower levels of anxiety and greater emotional resilience. The researchers concluded that humour serves as a buffer, protecting mental health during difficult times.

Laughter as Connection: Humour in Shared Hardship

One of humour's greatest gifts is its ability to connect people in adversity. Shared laughter creates solidarity, reminding us that we're not alone in our struggles.

Disaster Relief Teams
Relief workers responding to natural disasters often use humour to bond and maintain morale. A firefighter who worked during Hurricane Katrina recalled how his team joked about the absurdity of rescuing a cat stranded on a rooftop. "It wasn't about the cat," he said. "It was about giving ourselves something to laugh about so we could keep going."

Humour in Crisis Response
During the global refugee crisis, aid workers in Greece reported that humour often emerged in their interactions with displaced families. One refugee joked, "I walked so far I should be in the Olympics," eliciting laughter from everyone present. The shared humour bridged cultural differences and provided a moment of relief.

Laughing Through Tears

Humour in adversity is not about dismissing pain—it's about transcending it. From the trenches of war to the isolation of a pandemic, laughter becomes a way to reclaim humanity in the face of hardship. It connects us, lightens our burdens, and reminds us that even in the darkest times, joy is still possible.

As Viktor Frankl, a Holocaust survivor and psychiatrist, once wrote, "Humour is another of the soul's weapons in the fight for self-preservation." Laughter doesn't erase suffering, but it transforms it, turning tears into resilience and tragedy into hope.

CHAPTER 17: THE SATIRICAL SWORD

Humour As A Tool For Social Change

Humour is humanity's mirror, capable of reflecting not just our virtues but our flaws, hypocrisies, and injustices. Among the many forms of humour, satire stands out as the sharpest blade—a weapon wielded to critique authority, expose absurdities, and challenge societal norms. It is a rare blend of entertainment and provocation, forcing audiences to laugh and think simultaneously.

This chapter dives deeper into the dual-edged nature of satire, examining its historical roots, its evolution in contemporary culture, and its enduring power to spark dialogue and social change. Through expanded examples and fresh insights, we'll uncover how satire turns laughter into action, even as it treads the delicate line between critique and harm.

The Roots of Satire: Laughter as Subversion

Satire is as old as civilization itself, emerging wherever

humour has collided with power. In ancient Greece and Rome, playwrights and poets used satire to critique their societies. Aristophanes, often called the Father of Satire, targeted politicians, philosophers, and even gods with biting humour. His play Lysistrata, for example, ridiculed the absurdities of war by imagining women refusing their husbands sex until peace was negotiated.

Roman Satire: Horace and Juvenal
Roman satirists like Horace and Juvenal refined the art, balancing critique with elegance. Horace's satire was gentle and self-deprecating, using wit to inspire reflection without alienation. Juvenal, on the other hand, wielded satire with fiery indignation, exposing corruption and moral decay with lines like, "Who will guard the guards themselves?"

Juvenal's Enduring Relevance
Juvenal's biting phrase, panem et circenses (bread and circuses), remains a shorthand critique of societies that distract citizens with entertainment instead of addressing systemic problems. His words resonate in modern debates about media consumption, demonstrating satire's timeless ability to critique human priorities.

Swift's Masterstroke: The Precision of "A Modest Proposal"

No discussion of satire's power is complete without Jonathan Swift's A Modest Proposal. Published in 1729, the essay suggested that impoverished Irish families could sell their children as food for the wealthy, offering economic benefits while solving overpopulation.

Why It Worked

Swift's brilliance lay in his ability to present the grotesque with cold, rational logic. By mimicking the tone of an economist, he lured readers into his argument before shocking them with its inhumanity. The absurdity of the solution underscored the cruelty of the actual policies that dehumanized Ireland's poor.

Impact and Legacy

Swift's satire became a landmark in social critique, inspiring future writers to tackle injustice with humour. Though it did not immediately change British policies, it seeded awareness and remains a blueprint for how satire can provoke change.

"Vision is the art of seeing what is invisible to others," Swift wrote, a fitting description of how satire reveals truths hidden beneath the surface.

Satire Under Threat: The Cost of Speaking Truth

Satirists throughout history have faced persecution, imprisonment, or worse for daring to speak truth to power. In the modern era, the attack on Charlie Hebdo epitomized the risks of provocative satire.

The Controversy of Charlie Hebdo

Charlie Hebdo's cartoons often targeted religion, politics, and cultural taboos. Critics accused the magazine of insensitivity, while defenders argued that its work was essential to free speech. The terrorist attack in 2015, which killed twelve staff members, thrust the magazine into the global spotlight.

The Aftermath and Legacy

In the wake of the attack, Charlie Hebdo became a symbol of the precarious balance between satire, free expression, and respect for cultural boundaries. The incident highlighted satire's power to provoke—and the dangers of wielding that power in a polarized world.

Global Reactions

The slogan "Je Suis Charlie" became a rallying cry for free speech advocates worldwide. However, debates about the ethics of satirical provocation persisted, demonstrating the complex role satire plays in addressing sensitive issues.

Modern Satire: From Late Night to Social Media

Satire has found new platforms in the digital age, from the polished critiques of late-night television to the rapid-fire humour of memes. Each medium brings its own flavour to the tradition, reaching broader audiences than ever before.

Late Night's Sharp Tongue

Shows like The Daily Show and Last Week Tonight blend satire with investigative journalism, dissecting politics and culture with humour and intellect.

John Oliver's Net Neutrality Crusade

In 2014, John Oliver dedicated an entire segment to net neutrality, a complex and often-overlooked issue. His witty, accessible breakdown led to over 4 million public comments submitted to the FCC, demonstrating satire's ability to mobilize action.

Memes: Satire for the Digital Age

Memes distil complex ideas into digestible visuals, making them a powerful tool for critique. During the 2020 U.S. presidential election, memes lampooning both candidates spread rapidly, shaping public perception through humour.

Bernie Sanders and the Mittens

The viral image of Bernie Sanders at President Biden's inauguration became a global sensation, spawning thousands of memes. Beyond the humour, the image sparked discussions about wealth disparity and political elitism, showcasing the subtle power of internet satire.

Satire and Social Movements

Satire has long been a voice for the voiceless, amplifying marginalized perspectives and challenging oppressive systems. From feminist zines to LGBTQ+ comedians, humour has become a rallying cry for change.

Feminist Satire

Feminist writers and performers use satire to critique patriarchy and gender norms. Comedian Hannah Gadsby's Nanette deconstructed comedy itself, blending humour with raw vulnerability to challenge societal expectations of women and performers.

Queer Humour as Resistance

LGBTQ+ comedians like Wanda Sykes and Eddie Izzard use humour to expose homophobia and celebrate queer identity. Their work transforms laughter into solidarity, empowering communities through shared experiences.

Stonewall and Camp Humour

During the Stonewall riots of 1969, participants used campy humour to mock police and defuse tension. This playful defiance became a hallmark of queer resistance, demonstrating humour's ability to turn oppression into empowerment.

The Ethical Tightrope of Satire

Satire walks a fine line between critique and harm. While it has the power to expose injustice, poorly executed satire risks reinforcing stereotypes or alienating audiences.

Punching Up vs. Punching Down

Good satire "punches up," targeting those in power, rather than "punching down" on marginalized groups. When satire fails to recognize this distinction, it risks becoming part of the problem rather than the solution.

Case Study: Controversial Cartoons

A satirical cartoon intended to critique wealth inequality instead depicted a racial stereotype, sparking backlash. The controversy highlighted the importance of intent, execution, and audience context in crafting effective satire.

As Oscar Wilde noted, "If you want to tell people the truth, make them laugh, otherwise they'll kill you." Satire's power lies in its ability to soften hard truths with humour, but it requires care and precision.

The Sword That Cuts Both Ways

Satire is a sword, sharp and unforgiving. When wielded wisely, it carves paths to truth, exposing injustice and inspiring change. When misused, it alienates, divides, or wounds. From Swift's biting prose to the digital memes of today, satire remains one of humanity's most potent tools for holding power accountable and sparking transformation.

In George Orwell's words, "Every joke is a tiny revolution." Satire may not always topple regimes or rewrite policies, but it plants seeds of dissent, courage, and awareness. In a world full of noise, satire ensures that the truth—however uncomfortable—gets heard.

CHAPTER 18: COMEDY IN THE TRENCHES

Humour Amid Conflict

War is humanity stripped bare—a raw and unfiltered confrontation with mortality, chaos, and fear. Yet, paradoxically, war is also a crucible for some of the sharpest humour ever forged. Whether whispered between soldiers crouched in trenches, shared over meagre rations in prisoner-of-war camps, or exchanged by civilians in bomb shelters, humour emerges as an unlikely ally. It is a rebellion against despair, a communal thread that stitches fractured spirits together.

This chapter expands further into the role of humour amid conflict, offering additional real-life stories, cultural contexts, and psychological insights into how laughter becomes a survival mechanism in war zones and under oppression.

World War I: The Birth of Trench Humour

The Great War, with its seemingly endless stalemates and devastating losses, was fertile ground for what came to be known as trench humour. Soldiers lived for months, even years, in muddy, rat-infested trenches, where fear of death was constant. In these grim conditions, humour became both a coping mechanism and an act of camaraderie.

Letters from the Trenches
Trench humour often surfaced in letters sent home. Soldiers wrote absurd anecdotes that masked the horror of their daily lives but offered their families a glimpse of their resilience.

The "Monster Rats"
One British soldier described the rats in the trenches as "the size of cats, and twice as brave." In one letter, he wrote: "Last night, one of the chaps tried to shoot a rat with his revolver. Missed the rat but hit his own foot—guess who won that battle!" The humour wasn't just for show—it helped both soldiers and their loved ones endure the emotional strain of separation and uncertainty.

Slang and Nicknames
Trench slang was another outlet for humour. German artillery shells were nicknamed "Jack Johnsons" after the famous heavyweight boxer, while the endless mud was often referred to as "liquid real estate." These nicknames turned terrifying realities into something more manageable, allowing soldiers to laugh at what might otherwise paralyze them.

World War II: Humour as Unity

The Second World War brought the global scale of conflict to new extremes, yet it also amplified the role of humour as a unifying force.

Field Notes of Laughter

Frontline soldiers found ways to inject humour into their routines, even during active combat. American GIs stationed in the Pacific theatre often joked about their rations, especially the infamous canned meat product, SPAM.

SPAM's Comic Legacy

SPAM became a comedic icon in WWII. Soldiers created mock recipes like "SPAM Wellington" or "SPAM soufflé" and joked that it was the only food capable of surviving nuclear war. Years later, this humour would inspire Monty Python's iconic SPAM sketch, proving that wartime jokes have a way of outlasting the conflicts that birth them.

Comedians on the Frontline

During WWII, entertainers like Bob Hope brought laughter directly to the troops through USO tours. Hope's sharp one-liners and relatable humour became legendary, providing a rare moment of levity for soldiers far from home.

Bob Hope's Quip

During a performance for troops stationed in North Africa, Hope joked, "I've just come from Hollywood, where they're making a war movie. The battle scenes were so realistic that half the audience enlisted, and the other half ran for cover." The laughter wasn't just a reaction—it was a release, a momentary escape from the brutality of war.

POW Camps: Humour Behind Barbed Wire

Prisoner-of-war camps were spaces of profound suffering, but humour often emerged as a weapon of psychological survival. POWs created secret newspapers, staged comedic performances, and even crafted fake propaganda to undermine their captors.

The Great Escape: A Subversive Laugh

The Allied prisoners at Stalag Luft III, a German POW camp, famously orchestrated a massive escape attempt later immortalized in the film The Great Escape. But even before the escape, prisoners used humour as a form of subtle rebellion.

The "Goons"

POWs referred to German guards as "goons," a nickname that, while not overtly hostile, allowed them to strip their captors of their authority. One POW joked, "The goons think they're in charge, but we're running the real show—beneath their noses, literally." Such humour gave prisoners a sense of agency in an otherwise powerless situation.

Impromptu Theatre

POWs often staged theatrical performances that included comedic sketches mocking their captors. These shows weren't just entertainment; they were acts of psychological resistance, reminding the prisoners—and their captors—that humour could transcend oppression.

Civilian Resilience: Humour on the Home Front

War doesn't only affect soldiers—it reshapes civilian life in profound ways. For those living under air raids or in occupied territories, humour became a way to cope with fear and maintain a sense of normalcy.

The British Blitz Spirit
During the Blitz, Londoners endured nightly bombings that destroyed homes and claimed countless lives. Yet humour flourished amid the rubble.

Bomb Shelter Banter
In one bomb shelter, a man reportedly quipped, "The good news is, if they hit my house, I won't have to pay the mortgage!" The humour was dark, but it reflected the quintessentially British "stiff upper lip" that helped civilians endure the relentless attacks.

Occupied France: Humour as Resistance
In Nazi-occupied France, humour became a subtle form of defiance. Underground newspapers published satirical cartoons mocking German propaganda, while café patrons exchanged anti-Nazi jokes in hushed tones.

The French Café Whisper
One popular joke in occupied Paris went: "Why do the Germans wear steel helmets? Because they have to carry all the lies in their heads!" These jokes carried risks, but they reminded people of their shared resistance and humanity.

Humour's Role in Mental Fortitude

The psychological benefits of humour in wartime are profound. Laughter triggers the release of endorphins, reduces stress hormones, and fosters social bonds—all of which are crucial in high-stress environments.

The Science of Resilience
A 2015 study published in Psychological Science found that individuals who used humour to cope with trauma were more likely to maintain emotional resilience. The study highlighted humour's ability to reframe adversity, turning overwhelming situations into manageable narratives.

Case Study: Veterans' Laughter Groups
In recent years, veterans' organizations have embraced humour therapy as part of PTSD treatment. Group sessions often involve sharing war stories with a comedic twist, helping veterans process their experiences in a supportive, light-hearted environment.

Cultural Variations in Wartime Humour

Humour during conflict is universal, but its forms and expressions vary across cultures. In Japan during WWII, humour took on a quieter, more introspective tone, while in the Soviet Union, it was often laced with biting sarcasm aimed at both enemies and internal bureaucracy.

Soviet Humour: The Red Tape of War
Soviet soldiers frequently joked about the inefficiency of their own command structures. One common quip went, "The Germans might kill us, but our paperwork will bury us first." This dark humour served as both critique and camaraderie in a highly regimented system.

The Timelessness of Wartime Laughter

From ancient battles to modern-day conflicts, humour endures as a testament to human resilience. Whether whispered in trenches, scribbled in letters, or shared over rations, wartime laughter reminds us that even in the face of death, life finds a way to laugh.

As Charlie Chaplin once observed, "To truly laugh, you must be able to take your pain and play with it." In war, humour isn't just play—it's survival, a weapon against despair.

Laughter as a Quiet Victory

Comedy in the trenches is more than comic relief—it's a quiet victory against the chaos of conflict. From the biting quips of soldiers to the whispered jokes of civilians, humour becomes an act of courage, defying fear with joy. It reminds us that even in the darkest moments, laughter can illuminate a path forward.

CHAPTER 19: LAUGHING AT FEAR

The Role Of Dark Humour

Dark humour dances on the edge of taboo, playing with themes that others may shy away from. It transforms mortality, disaster, and existential dread into punchlines, creating a space where laughter coexists with discomfort. To some, it might seem irreverent or inappropriate. But to those in the throes of crises—first responders, healthcare workers, and communities facing trauma—it is a lifeline, a shield, and even a weapon.

This further expansion of Chapter 19 delves deeper into the nuances of dark humour, exploring additional studies, historical and cultural examples, and its impact across different spheres of human experience.

The Anatomy of Dark Humour: Why We Laugh in the Face of Fear

Dark humour often starts where traditional jokes end: in the space of discomfort. It thrives on our collective discomfort with mortality, unmasking the inevitable and making us laugh at what we cannot control. It isn't about trivializing pain; it's about defying it.

The Cognitive Alchemy of Dark Humour

Dark humour works through cognitive dissonance. It forces the brain to reconcile two seemingly contradictory ideas—fear and amusement—creating a release of tension that feels both profound and rebellious.

Study Insight: A 2017 study in Cognitive Processing found that individuals with a high appreciation for dark humour tend to have above-average intelligence and emotional stability. Researchers theorize that processing dark humour requires mental agility to navigate its complexity and emotional resilience to embrace its audacity.

The Emergency Room Banter

An ER doctor recalled a particularly chaotic night when a nurse quipped, "At this rate, we're going to need a 'Buy One, Get One Free' coupon for stitches." The joke wasn't at the expense of the patients—it was a shared coping mechanism that diffused the tension in the room, helping the team stay focused and composed.

Dark Humour in Crisis Professions: The Brave and the Broken

First responders, firefighters, police officers, and healthcare workers often inhabit a world where life and death hang in delicate balance. Their use of dark humour isn't an act of cruelty—it's a form of psychological armour.

The Firefighter's Philosophy

Firehouses are famous for their gallows humour. Whether joking about the chief's cooking or the inevitable mess of a call, the humour serves to build bonds and lighten the emotional load.

The Helmet Story
A firefighter recalled pulling a cat out of a tree during his first week on the job. His seasoned colleagues joked, "Don't let this go to your head—it's all downhill from here." While seemingly light-hearted, the humour prepared him for the grim realities ahead, balancing levity with pragmatism.

Law Enforcement: The Thin Blue Laugh
Police officers face immense stress and danger daily. Their humour often reflects the absurdity of their situations, with jokes aimed not at victims but at the sheer unpredictability of their work.

The Traffic Stop Quip
During a routine stop, an officer was asked why he carried two sets of handcuffs. He replied, "One for the criminal, the other for me when I hear their excuse." The humour wasn't dismissive—it acknowledged the tension of their work while softening the interaction.

Healthcare Workers: Jokes on the Edge of Mortality

Healthcare settings are particularly fertile ground for dark humour, as professionals navigate not only the fragility of life but also the emotional toll of their work.

Hospice Care and Shared Laughter
In end-of-life care, humour often becomes a bridge between patient and caregiver. A terminally ill man,

while discussing his final wishes, told his hospice nurse, "Don't let them bury me in that ugly brown suit—I've got better taste than that." The laughter that followed wasn't at the expense of the moment but a poignant reminder of his individuality and humanity.

The Coffee Machine Saga
One hospital's coffee machine broke down during a hectic shift. A nurse taped a sign to it that read, "RIP—cause of death: overwork." The joke became a running gag in the unit, turning frustration into a shared moment of levity.

Historical Context: Dark Humour as Resistance

Throughout history, dark humour has been a tool of resistance, particularly in times of oppression and war. It allows communities to assert their humanity, mocking their oppressors or circumstances even when they seem insurmountable.

Jewish Humour During the Holocaust
Even in concentration camps, humour persisted as an act of spiritual rebellion. Prisoners told jokes that undermined Nazi propaganda, reclaiming a sense of agency in the face of unimaginable horror.

The Barbed Wire Joke
One joke went: "In the camps, we get plenty of exercise. Morning stretches to avoid bullets, afternoon jogs to the workstations, and evening climbs over barbed wire—if you're lucky." The humour wasn't cruel; it was a way to confront despair with courage.

Cold War Quips: Soviet Sarcasm
During the Cold War, Soviet citizens used dark humour

to critique their government under the guise of jokes. One popular quip was, "The future is certain; it's the past that's always changing." This humour didn't change their reality, but it gave voice to collective frustration.

Cultural Perspectives: How Dark Humour Varies Globally

Dark humour may be universal, but its expressions are deeply influenced by cultural norms and histories.

Scandinavian Humour: Dry and Existential
Scandinavian humour often leans into existentialism, reflecting the region's cultural grappling with life's big questions. A popular Norwegian joke goes, "How do you start a Viking funeral? With a lot of paperwork." This humour embraces the region's bureaucracy and dark winters with wry self-awareness.

Irish Humour: Laughing at Suffering
Ireland's history of famine, colonization, and conflict has cultivated a tradition of darkly comedic storytelling. Irish funeral humour is particularly notable, blending grief with laughter in a way that honours both.

The Wake Toast
At an Irish wake, a mourner raised a glass and said, "Here's to Mike. He finally found a way to beat the taxman." The laughter didn't diminish the loss; it celebrated the life and wit of the deceased.

Dark Humour in Modern Media: Entertainment with an Edge

Contemporary media has embraced dark humour, using

it to tackle existential themes with sharp wit.

Television and Streaming
Shows like Fleabag, BoJack Horseman, and The Office frequently use dark humour to explore themes of loss, failure, and self-doubt. These shows resonate with audiences because they acknowledge life's difficulties while finding humour in them.

Memes: The Collective Catharsis
The internet has democratized dark humour, with memes providing a global platform for collective catharsis. During the pandemic, memes about isolation and uncertainty offered a shared outlet for frustration and fear.

The Ethical Dilemmas of Dark Humour

Dark humour's strength is also its greatest risk: poorly executed jokes can alienate or harm, especially when they target vulnerable groups.

Navigating the Line
Effective dark humour "punches up," targeting systems of power or shared existential fears rather than individuals or marginalized communities. The line between clever and cruel often lies in intent and context.

The Cancelled Comedy Show
A comedian's joke about a recent disaster went viral for all the wrong reasons. While some defended it as satire, the backlash underscored the importance of sensitivity in navigating tragedy with humour.

Laughing Into the Void

Dark humour isn't for everyone, but for those who embrace it, it offers something profound—a way to laugh at the absurdities of life, death, and everything in between. It is defiant and raw, a reminder that even when the world feels unbearable, we can still find reasons to smile.

In the words of Kurt Vonnegut, "Laughter and tears are both responses to frustration and exhaustion. I myself prefer to laugh, since there is less cleaning up to do afterward." Dark humour may not erase fear, but it transforms it, turning shadows into a canvas for courage and connection.

CHAPTER 20: HUMOUR AS A HEALING BALM

The Role Of Laughter In Medicine

Laughter is a paradox: both instinctive and complex, spontaneous yet deliberate. In medicine, it has emerged not just as a diversion from pain but as a complementary therapy, weaving humour into the tapestry of healing. It lightens heavy moments, soothes emotional wounds, and even accelerates physical recovery. This chapter further expands on the role of humour in medicine, offering new insights, anecdotes, and studies to illustrate how laughter truly is a balm for the human condition.

Laughter's Biochemical Symphony: The Physiology of Healing

Laughter isn't just an outward expression; it triggers a cascade of internal processes that promote healing. The moment we laugh, the brain releases a cocktail of endorphins, serotonin, and dopamine, countering the effects of stress hormones like cortisol. This biochemical response relaxes the body, reduces pain perception, and even strengthens the immune system.

Study Insight: Laughter and Heart Health
A study conducted at the University of Maryland found that laughter improves vascular function. Participants who watched a comedy film experienced significant blood vessel dilation, improving circulation and reducing cardiovascular risk. The researchers concluded that 15 minutes of laughter a day could have similar benefits to aerobic exercise.

The "Comedy Prescriptions"
At a cardiac clinic in London, doctors encouraged patients to watch stand-up comedy videos during recovery. One patient reported, "I laughed so hard at Eddie Izzard, I forgot I'd just had surgery!" His recovery progressed faster than expected, illustrating the tangible impact of humour on healing.

Laughter as an Analgesic: Turning Pain into Punchlines

Pain is one of medicine's greatest challenges, but humour offers a unique way to address it. By distracting the mind and activating the body's natural painkillers, laughter reduces discomfort and enhances quality of life.

Case Study: Burn Patients and Humour Therapy
A 2011 study in Burns journal explored the effects of humour therapy on burn patients undergoing painful wound care. Patients who were shown humorous videos during dressing changes reported significantly lower pain levels compared to those who did not. Laughter,

in this case, acted as a mental aesthetic, reframing the experience.

The "Pain-Free Zone"
In a children's burn unit in Boston, a nurse nicknamed their humour-filled ward "The Pain-Free Zone." Staff members dressed as superheroes or clowns, creating a playful atmosphere that helped young patients endure painful treatments. One child, after being asked if he needed more medicine, replied, "Nope, just send in Batman!"

Norman Cousins: A Pioneer in Laughter Therapy

Norman Cousins' groundbreaking recovery story remains one of the most compelling examples of humour's healing power. Diagnosed with ankylosing spondylitis, a degenerative condition, Cousins rejected the prognosis that he had only months to live. Instead, he crafted his own recovery plan, which included massive doses of vitamin C and daily doses of laughter through comedy films.

The 10-Minute Rule
Cousins observed that ten minutes of belly laughter would give him two hours of pain-free sleep. By immersing himself in the humour of Charlie Chaplin and the Marx Brothers, he rewired his brain's response to pain. His condition improved so dramatically that he returned to work within months.

Legacy: Cousins' story inspired a wave of research into psychoneuroimmunology—the study of how mental states influence physical health—and cemented laughter's role as a legitimate medical tool.

Clown Doctors: Humour in Paediatric Care

For children in hospitals, the unfamiliarity of medical procedures and the sterile environment can be terrifying. Clown doctors—trained performers who blend humour, music, and magic—turn these fears into moments of joy.

Clowning in Crisis
In a paediatric oncology unit in Brazil, clown doctors performed weekly visits, engaging children with slapstick antics and light-hearted jokes. One child undergoing chemotherapy said, "They make the medicine taste like candy!" The program not only reduced the children's anxiety but also improved cooperation during treatments.

Study Insight: A 2013 study in The Journal of Health Psychology found that clown interventions significantly reduced cortisol levels in hospitalized children, indicating lower stress levels.

Humour for Caregivers: Easing the Emotional Load

Caregivers—whether healthcare professionals or family members—often bear the emotional weight of illness. Humour helps them navigate the complexities of caregiving, offering relief from the stress and burnout that can accompany their roles.

Caregiver Comedy Nights
In a California hospice program, a support group for caregivers organized monthly comedy nights

where participants shared funny anecdotes about their experiences. One caregiver recounted how her father, after refusing to take his medication, quipped, "If I'm going to the pearly gates, I'd better look sharp!" The laughter that followed was cathartic, reminding caregivers of the joy amid the sorrow.

Comedian George Burns once said, "Happiness is having a large, loving, caring, close-knit family in another city." For caregivers, humour often provides a much-needed escape, lightening the emotional load.

Humour in the Final Chapter: Hospice and End-of-Life Care

In hospice settings, humour becomes a bridge between the inevitable and the meaningful. It allows patients and families to face mortality with dignity, laughter, and even joy.

A Terminally Funny Patient
A terminally ill patient in a hospice centre requested a prank funeral rehearsal, complete with fake eulogies. One family member jokingly declared, "We'll miss your cooking—but not your burnt toast!" The laughter that filled the room turned what could have been a sombre moment into a celebration of the patient's life.

Study Insight: Humour as Emotional Support
A 2018 study in Palliative & Supportive Care found that hospice patients who engaged in humour-based activities reported greater satisfaction with their care and improved relationships with their families. The study concluded that humour fosters emotional intimacy,

easing the transition during end-of-life care.

Post-Trauma Healing: Laughter in Disaster Recovery

Communities recovering from natural disasters or collective trauma often use humour to rebuild morale and strengthen bonds.

Hurricane Harvey and the Resilient Spirit
After Hurricane Harvey devastated Houston, a local woman posted a viral video of herself rowing a kayak through her flooded living room while wearing a party hat. She joked, "All I'm missing is the piña colada!" The humour offered a moment of levity amid the chaos, reflecting the resilience of the human spirit.

The Role of Technology: Laughter at Your Fingertips

The digital age has brought humour therapy to new heights, with apps, virtual clown visits, and telemedicine integrating laughter into patient care.

Humour Apps for Patients
Apps like Laughly provide curated stand-up comedy routines tailored to different moods. Hospitals have begun recommending these apps to patients as part of their wellness programs, offering an on-demand dose of laughter.

A Virtual Clown
During the COVID-19 pandemic, hospitals in Italy used video calls to connect paediatric patients with clowns. One child, speaking with a clown on-screen, joked, "I'll trade you my pudding cup if you show me another magic

trick!" The interaction brought smiles to both the child and the medical staff.

Laughter as Medicine

Humour in medicine is more than just a distraction—it is a transformative force. It reduces pain, alleviates stress, builds resilience, and fosters connection between patients, caregivers, and medical professionals. As technology advances and research deepens, the integration of humour into healthcare will only grow, proving that laughter truly is the best medicine.

In the words of Maya Angelou, "I don't trust anyone who doesn't laugh." In medicine, laughter is the ultimate trust-builder, a reminder that even in the most challenging times, there is always room for joy.

CHAPTER 21: CULTURAL LAUGHTER

Humour Across The Globe

Humour is the heartbeat of culture, a vibrant expression of human creativity and connection. It bridges gaps, unearths truths, and reveals the soul of a society. While laughter itself is universal, what makes people laugh is deeply rooted in their history, values, and way of life. To dive deeper into the intricacies of global humour is to uncover the intimate and diverse ways cultures navigate their joys, sorrows, and absurdities.

In this expanded exploration, we examine humour not just as entertainment but as a profound cultural artifact, diving into its deeper societal implications and offering further real-life examples to illustrate its nuances.

The Language of Humour: A Linguistic Playground

Language is both the vehicle and the playground of humour, shaping its delivery and reception in fascinating ways. Wordplay, one of the most ancient forms of humour, relies heavily on linguistic structure and cultural context.

Japanese Kakekotoba: The Pivot of Wit
In Japan, kakekotoba (pivot words) are a sophisticated form of punning that play on the dual meanings of words. Unlike direct jokes, this humour is subtle, rewarding those who grasp its linguistic layers.

Deeper Example: Kakekotoba in Poetry
In traditional Japanese waka poetry, a line might use the word matsu to mean both "pine tree" and "to wait." For example, a poet might write, "I wait beneath the pines for your return," blending natural imagery with emotional longing. While not overtly comedic, this linguistic elegance reflects a humour rooted in intellect and creativity.

Arabic Humour: Exaggeration and Metaphor
Arabic humour often uses exaggeration and metaphor to make a point or deliver a punchline. Stories and jokes are frequently built around elaborate scenarios, with humour emerging from the unexpected twists.

The Tailor's Promise
An old Arabic joke tells of a tailor who promises a customer, "Your suit will be ready in three days, just like God created the heavens and the earth." When the customer returns to find the suit unfinished, the tailor shrugs and says, "But look at the state of the world—some jobs are never really done!" This blend of metaphor and wry commentary captures the wit of everyday life in the Arab world.

Humour as a Cultural Compass: What Societies Value

Humour reflects the values, struggles, and triumphs of the cultures it arises from. What a society finds funny often reveals its priorities and tensions.

German Precision and Absurdity

While Germans are often stereotyped as humourless, their comedic traditions—particularly in cabaret and satire—showcase a sharp and sometimes absurd wit. German humour often mocks bureaucracy and the rigidity of rules.

Deeper Example: The Kafkaesque Joke

A common German joke goes: "A man goes to the city office to change his address. After hours of paperwork, the clerk says, 'Your new address is the same as your old one.' The man replies, 'Yes, but at least now it's official.'" The humour lies in the absurdity of excessive bureaucracy, a recurring theme in German satire.

Indian Humour: Chaos and Colour

India's humour reflects its vast diversity, often embracing the chaos of life with vibrant wit. Jokes about overbearing relatives, crowded trains, or elaborate wedding traditions resonate across the country.

The Wedding Mishap

At a wedding in Mumbai, the groom joked to his bride during the ceremony, "If you see my aunt coming with more sweets, run!" The quip, shared in front of a lively crowd, poked fun at the cultural tendency to overindulge guests, highlighting India's humour in everyday excess.

Social Commentary Through Humour: Laughing at Authority

Humour often serves as a form of resistance, a way to critique authority or highlight societal flaws without direct confrontation. This tendency manifests differently across cultures.

French Satire: Wit with an Edge

France has a rich tradition of satire, blending intellect with provocation. From Voltaire's sharp critiques to modern satirical publications like Charlie Hebdo, French humour often pushes boundaries to provoke thought and dialogue.

Deeper Example: Voltaire's Lampoon

In his novella Candide, Voltaire mocks the philosophical optimism of his time with biting humour. When Candide declares, "All is for the best in the best of all possible worlds," the absurdity of his endless misfortunes underscores the folly of blind idealism. This brand of satire has shaped French humour as both intellectual and daring.

African Humour: Resilience and Joy

Across African cultures, humour often functions as a communal experience, addressing social inequities with a mix of laughter and wisdom. Jokes and storytelling are frequently used to impart lessons or defuse tension.

The Village Politician

In a Kenyan village, a local politician promised, "I will fix all the roads before the next rainy season!" When the rains came and the roads remained muddy, villagers began referring to potholes as "campaign promises." The humour turned frustration into shared resilience, using wit to hold power accountable.

Global Celebrations of Humour: Festivals of Laughter

Around the world, festivals and traditions often incorporate humour as a central element, showcasing its ability to unite communities.

Scotland's Hogmanay and Humour

During Hogmanay, Scotland's New Year celebration, humour takes centre stage in street performances and storytelling. Comedians often weave jokes about the weather into their acts, with one performer quipping, "Our summer lasted all of yesterday—it was glorious!"

Mexico's Día de los Santos Inocentes

Mexico's "Day of the Holy Innocents," similar to April Fools' Day, features pranks and playful teasing. The tradition reflects the Mexican spirit of chingón—resilience through laughter, even in the face of hardship.

A Family's Practical Joke

During Día de los Santos Inocentes, a mother replaced her family's sugar with salt for their morning coffee. The ensuing laughter turned the prank into a cherished memory, reflecting the warmth of Mexican humour.

Humour Lost and Found in Translation

Cross-cultural humour can lead to delightful misunderstandings or unintentional hilarity, as jokes don't always translate perfectly.

The Swedish Chef

The character of the Swedish Chef on The Muppet Show became a global phenomenon, with his nonsensical "Swedish" accent. While Swedes initially found the portrayal baffling, they eventually embraced the character, laughing along with the world at the affectionate parody.

Humour as a Universal Bridge

Despite its cultural nuances, humour transcends borders, reminding us of our shared humanity. A good joke about love, family, or life's absurdities can resonate across languages, creating connections where words might fail.

As George Bernard Shaw remarked, "The single biggest problem in communication is the illusion that it has taken place." Humour bridges this gap, ensuring that the message is not only delivered but felt.

A World of Laughter

Cultural humour is a kaleidoscope, reflecting the intricate patterns of human society. Whether subtle or bold, dry or exuberant, humour captures the essence of what it means to be human. Across the globe, laughter reminds us of our similarities even as it celebrates our differences. In a world often divided, it is perhaps the most universal of languages.

CHAPTER 22: FAITH AND FOLLY

Humour In Religion

Religion, like humour, addresses the deepest aspects of human existence—life, death, love, and the mysteries of the divine. While some may see faith and folly as opposites, they are often partners in reflection. Humour in religion serves as a mirror, reflecting human frailty and imperfection while illuminating spiritual truths in ways that are accessible and memorable.

In this expanded exploration, we'll dive further into the coexistence of humour and reverence across traditions and contexts, enriching our understanding with more real-life examples, quotes, and cultural insights.

Humour in Sacred Texts: Hidden Gems of Laughter

Humour in religious scriptures often takes the form of irony, exaggeration, or absurdity, serving to teach moral lessons or reveal human flaws.

Jewish Scripture: Abraham's Negotiation
The Book of Genesis contains an often-overlooked

moment of humour when Abraham bargains with God over the fate of Sodom and Gomorrah. Abraham starts by asking if God would spare the city for fifty righteous people, then whittles the number down to ten. The exchange is both bold and humorous, showing Abraham's tenacity and God's willingness to engage.

As Rabbi Harold Kushner noted, "The Bible is filled with humour because it understands that laughter is as divine as tears."

Hindu Scriptures: The Cosmic Joke
In Hinduism, humour often appears in myths and epics, reflecting the playfulness of the divine. Lord Krishna, a central figure in Hindu texts, is renowned for his mischievous pranks, from stealing butter to playfully teasing his devotees. These stories celebrate divine joy and remind followers not to take life too seriously.

Krishna and the Butter
One tale recounts how Krishna stole butter from his mother's storeroom, only to charm his way out of punishment with a mischievous smile. The story, retold in countless festivals, reflects the balance between devotion and light-heartedness in Hindu spirituality.

Christian Parables: The Irony of the Pharisees
Jesus's parables in the New Testament often contain moments of irony and humour. When the Pharisees criticize him for dining with sinners, Jesus replies, "It is not the healthy who need a doctor, but the sick." The pointed humour underscores the Pharisees' hypocrisy, using wit to deliver a profound spiritual message.

A Modern Take

In a contemporary sermon, a pastor retold this story, quipping, "If you're perfect, this church probably isn't for you—we're saving seats for the sinners." The humour brought the lesson to life, making it relatable to modern listeners.

Humour in Rituals and Community Practices

Religious rituals, though solemn at their core, often incorporate humour to foster connection and joy.

Buddhist Humour: Laughing Monks

In Buddhist monasteries, humour often punctuates teachings. Monks may playfully scold one another or tell jokes to illustrate spiritual truths. This light-heartedness reflects Buddhism's emphasis on mindfulness and the transient nature of life.

The Dalai Lama's Wit

The Dalai Lama is known for his humour. When asked by a journalist if he would reincarnate as a woman, he replied, "Only if she's attractive!" While light-hearted, his humour disarms and connects, reflecting his belief in compassion and relatability.

Catholic Laughter: Feast of Fools

In medieval Europe, the Feast of Fools allowed clergy to mock their own hierarchy. Lower-ranking members dressed as bishops and staged comedic parodies of church ceremonies. While irreverent, the festival provided a cathartic outlet for humour within the structure of the Church.

Impact: The Feast of Fools reflected the Church's recognition that laughter could coexist with reverence, offering relief from the rigidity of religious life.

Modern Religious Humour: Satire in an Age of Scepticism

Contemporary humour often uses religion as a lens to critique hypocrisy, encourage dialogue, or explore existential questions. While some satire is controversial, much of it reveals deep engagement with faith.

Jewish Stand-Up Comedy: Laughing Through Adversity
Jewish comedians like Mel Brooks and Joan Rivers often use humour to explore faith and identity. Brooks's film History of the World, Part I includes a scene where Moses accidentally drops a third tablet of commandments, leaving humanity with only ten. The humour is irreverent but affectionate, reflecting a tradition that embraces questioning.

As Mel Brooks quipped, "If God wanted us to be solemn, he wouldn't have made us funny."

Islamic Humour: A Gentle Laugh
While Islamic humour is often subtle, it reflects the deep humanity of the faith. Comedian Hasan Minhaj frequently explores the intersection of religion and culture, using humour to address stereotypes and misconceptions about Islam.

Minhaj's Wedding Joke
In one routine, Minhaj jokes about how his parents reacted when he chose a non-Muslim wife: "They asked if

she could cook biryani. I said no, but she makes a mean lasagna. They were halfway sold." The humour highlights the tension between tradition and modernity, fostering understanding through laughter.

Cultural Variations: How Humour and Faith Interact

The relationship between humour and religion varies widely across cultures, reflecting different attitudes toward reverence and play.

African Traditions: Humour as Resilience
In African spiritual practices, humour often appears in folktales where trickster figures like Anansi the Spider use wit to overcome challenges. These stories celebrate resilience, teaching moral lessons with a smile.

Anansi Outsmarts a Leopard
In one tale, Anansi convinces a leopard to trap itself in a net by pretending it's a new fashion. The humour underscores the value of intelligence over brute force, resonating with African values of community and cleverness.

Scandinavian Humour: The Sacred Absurd
In Nordic mythology, gods like Loki embody both chaos and humour, reflecting the Scandinavian appreciation for life's absurdities. Loki's mischief often leads to trouble, but it also reveals deeper truths about the fragility of existence.

Loki's Shapeshifting
One story recounts how Loki turned into a mare to distract a giant's horse, resulting in the birth of Sleipnir,

an eight-legged steed. The humour lies in Loki's audacity, but the tale also highlights the gods' fallibility—a central theme in Norse mythology.

Balancing Satire and Sensitivity

Humour in religion requires a delicate balance. While it can foster connection and critique, it can also offend when poorly executed or taken out of context.

Case Study: Religious Comedy in Film
The Coen Brothers' A Serious Man explores Jewish faith through dark humour, chronicling a man's struggle to find meaning amid life's absurdities. The film's humour is both existential and compassionate, inviting audiences to laugh and reflect simultaneously.

The Universality of Sacred Laughter

Across traditions, humour serves as a reminder that faith is not about perfection but about connection—to the divine, to others, and to oneself. It breaks down barriers, turning solemnity into joy and dogma into dialogue.

The Rabbi's Dancing Shoes
A rabbi, known for his stern demeanour, surprised his congregation by dancing joyfully during a festival. When asked why, he joked, "Even Moses needed a break from carrying the tablets!" The moment united the community in laughter, celebrating both faith and humanity.

Divine Laughter

Humour in religion is not a departure from reverence but a deepening of it. It reminds us that the divine is not distant or unapproachable but present in our imperfections and joys. Through humour, faith becomes not just a solemn duty but a vibrant celebration of life.

As the Sufi poet Rumi wrote, "When the soul lies down in that grass, the world is too full to talk about. Ideas, language, even the phrase each other doesn't make any sense." In that space, laughter bridges the sacred and the human.

CHAPTER 23: HUMOUR AS A LINGUISTIC PLAYGROUND

The Power Of Wordplay

Language is the ultimate playground for humour. Its twists and turns, ambiguities, and double meanings invite infinite possibilities for wit. Puns, riddles, and linguistic quirks create humour that transcends words, challenging our intellect and tickling our funny bone. Whether in Shakespeare's clever dialogues, cross-cultural jokes, or the dad jokes that inspire groans, wordplay is a universal testament to human creativity.

This expanded chapter dives even deeper into the mechanics, history, and cultural significance of wordplay, revealing how it shapes humour and reflects the rich complexities of human communication.

The Dual Nature of Wordplay: Cleverness and Connection

Wordplay operates on two levels. First, it delights through clever manipulation of language. Second, it fosters connection, inviting listeners to share in the humour of a well-crafted pun or riddle.

What Makes Wordplay Funny?
Linguistic humour often hinges on incongruity—the gap between expectation and outcome. When a pun subverts what we anticipate, our brain experiences a jolt of surprise, followed by amusement as we reconcile the ambiguity.

Study Insight: A 2020 study in NeuroImage showed that understanding wordplay activates the brain's language and reward centres simultaneously, creating a satisfying cognitive experience.

Example: A Timeless Pun
"I'm on a seafood diet. I see food, and I eat it." This pun plays on the dual interpretation of "seafood" and "see food," rewarding listeners for catching the wordplay.

Shakespeare: The Maestro of Linguistic Humour

William Shakespeare's plays are treasure troves of wordplay, weaving humour seamlessly into complex narratives. His puns are not merely comedic; they reveal character, deepen themes, and engage audiences across centuries.

Puns in Twelfth Night
In Twelfth Night, Maria says of Malvolio: "He does smile his face into more lines than is in the new map with the augmentation of the Indies." The joke combines wordplay

with satire, poking fun at both Malvolio's exaggerated expressions and the geopolitical complexities of the age.

Double Meanings in Hamlet

Hamlet's famous line to Polonius, "You are a fishmonger," carries layers of meaning. While it seems literal, "fishmonger" was Elizabethan slang for a pimp, adding a sharp edge to Hamlet's wit. The wordplay mocks Polonius's obsequiousness while hinting at his manipulative tendencies.

As Shakespeare wrote in As You Like It, "The fool doth think he is wise, but the wise man knows himself to be a fool." This line itself plays on linguistic inversion, demonstrating the bard's mastery of humour.

Riddles Through the Ages: Humour in the Unknown

Riddles are among the earliest forms of wordplay, blending humour with intellectual challenge. Their enduring appeal lies in their ability to amuse while demanding active engagement.

The Riddle of the Sphinx

In Greek mythology, the Sphinx poses a riddle to travellers: "What walks on four legs in the morning, two legs at noon, and three legs in the evening?" The answer—"a human"—combines metaphor with humour, encapsulating the stages of life.

Modern Riddles: Humour in Simplicity

"What comes once in a minute, twice in a moment, but never in a thousand years?"
Answer: The letter M. The humour lies in the misleading setup, which directs the listener toward a temporal

interpretation before revealing a linguistic twist.

Cultural Dimensions of Wordplay: The Joy and Challenge of Translation

Wordplay is deeply rooted in linguistic structure, making it challenging to translate. Yet, successful adaptations often reveal the universality of humour.

Japanese Dajare: Homophonic Humour
In Japanese, dajare jokes rely on homophones and word overlaps. For example:
"Why do cranes stand on one leg? Because if they lifted both, they'd fall over!" This joke translates well, preserving the humour in its logic, though many dajare lose their punch outside the Japanese language.

Arabic Linguistic Humour: Rich in Metaphor
Arabic jokes often use wordplay to critique societal norms or reflect cultural values. A classic Arabic pun goes: "A man asked his friend, 'How do you find life?' The friend replied, 'I open the door, and it's usually just sitting there.'" The humour plays on the double meaning of "find life," blending philosophy with absurdity.

Lost in Translation: A Chinese Example
In Mandarin, the phrase mei you qian de gou means "a dog with no money" but sounds identical to "a dog with no front teeth." A joke using this phrase depends on its auditory pun, which often baffles non-Chinese speakers unfamiliar with tonal nuances.

The Psychology of Puns: Groans and Gratification

Puns often elicit groans as much as laughter, yet even the groans indicate engagement. This dual reaction highlights the brain's love-hate relationship with linguistic humour.

The Groan Mechanism
When we hear a pun, our brain works to resolve its ambiguity. Even if the humour feels "cheap," the act of decoding it provides satisfaction, making puns a reliable form of amusement.

The Dad Joke
"Why don't skeletons fight each other? They don't have the guts." The simplicity of this joke illustrates why wordplay remains a staple of casual humour—it's accessible and easy to share.

Wordplay as Social Commentary

Beyond entertainment, wordplay often serves as a subtle form of critique or reflection.

French Wit: Double Entendres
French humour frequently employs double entendres, blending elegance with pointed critique. During the French Revolution, wordplay became a form of resistance. A satirical line from the period mocked the monarchy: "The king loses his head but keeps his crown." The pun critiques both literal and symbolic loss, illustrating the sharpness of revolutionary humour.

American Wordplay: Humour in Headlines
Newspaper headlines often use puns to grab attention. A memorable example: "Time Flies Like an Arrow;

Fruit Flies Like a Banana." The humour plays on the grammatical structure, challenging readers to parse its dual meanings.

Wordplay in Modern Media: New Frontiers

In the digital age, wordplay has adapted to new platforms, thriving in memes, advertising, and viral tweets.

Memes and Hashtags
Social media platforms like Twitter and Instagram amplify wordplay through hashtags and meme culture. During the height of the pandemic, the hashtag QuaranTEAn humourously promoted tea-drinking as a comforting activity, blending linguistic humour with social commentary.

Brand Slogans
Brands often use puns to create memorable slogans. For instance, a hair salon called "Curl Up and Dye" uses wordplay to grab attention and convey its services with humour.

The Universality of Wordplay: Laughter Beyond Language

While wordplay often relies on specific linguistic quirks, its essence transcends language. The delight in clever twists and unexpected meanings is a universal experience, reflecting humanity's shared joy in creativity.

International Wordplay Competition
The annual O. Henry Pun-Off in Texas attracts

participants from around the world, showcasing the global appeal of wordplay. One memorable entry: "I used to be a baker, but I couldn't make enough dough." The joke resonates universally, proving that linguistic humour knows no borders.

The Infinite Joy of Linguistic Play

Wordplay is more than a comedic device—it's a celebration of language itself. Whether through Shakespeare's layered puns, riddles that challenge the mind, or cross-cultural jokes that bridge divides, linguistic humour reveals the endless possibilities of human expression. In the world of wordplay, every twist and turn of language becomes a source of joy, surprise, and connection.

As linguist Victor Borge said, "Laughter is the shortest distance between two people." Wordplay, with its clever manipulation of language, shortens that distance further, creating bonds through the shared pleasure of discovery.

CHAPTER 24: THE FEMININE WIT

Gendered Perspectives On Humour

Humour, like all art forms, reflects societal structures and dynamics. For centuries, women's humour has both adapted to and challenged gender norms, serving as a subversive tool to address inequality, assert agency, and redefine cultural narratives. From the sharp wit of suffrage leaders to the unapologetic boldness of modern feminist comedians, women's humour is a powerful force that has evolved alongside their roles in society.

In this chapter, we delve into the history and significance of feminine wit, exploring how women have used humour to navigate oppression, claim visibility, and reshape the comedic landscape.

The Historical Foundations of Feminine Wit

Humour has long been a survival tool for women in patriarchal societies. Historically, women used wit as a way to challenge constraints subtly, navigating the thin line between rebellion and acceptability.

Humour in the Suffrage Movement

During the fight for women's right to vote, suffragettes used humour to deflate their opponents' arguments and rally support for their cause. Cartoons, speeches, and parodies showcased their sharp wit and resilience.

Suffrage Satire

A suffragette cartoon depicted a man crying at home, surrounded by children, while his wife votes. The caption read, "Who will mind the baby?"—a pointed critique of the double standards imposed on women's roles. The humour exposed the absurdity of fears surrounding women's political participation.

Dorothy Parker: The Queen of Literary Wit

Dorothy Parker, a prominent figure in the 1920s Algonquin Round Table, epitomized feminine humour with her biting one-liners. Known for her sardonic wit, she once quipped, "Men seldom make passes at girls who wear glasses." Her humour reflected the limitations of societal expectations while cleverly mocking them.

Feminist Comedy: Humour as Resistance

The feminist movements of the 20th and 21st centuries brought a resurgence of humour as a form of activism, challenging stereotypes and empowering women through laughter.

Trailblazers of Feminist Comedy

Comedians like Phyllis Diller and Joan Rivers paved the way for female voices in comedy. Diller used self-deprecating humour to critique societal pressures on women, joking, "The only thing domestic about me is that I was born in this country." Rivers, on the other hand, tackled taboo topics with unflinching boldness, breaking barriers for women in stand-up.

Joan Rivers on Ageism
Rivers once joked, "At my age, I don't buy green bananas." The humour, while light-hearted, highlighted society's obsession with youth and the challenges of aging, particularly for women.

The Second Wave: Humour Gets Political
In the 1970s, feminist comedians like Lily Tomlin and Elaine May infused their work with sharp political commentary. Tomlin's sketches often critiqued workplace inequality and gender roles, blending humour with incisive social critique.

Modern Example: Tina Fey and Amy Poehler
The duo's humour on Saturday Night Live and beyond has redefined feminist comedy. Their skit "Bitches Get Stuff Done" humourously reclaimed a derogatory term while celebrating women's resilience and competence.

Women of Colour in Comedy: Expanding the Landscape

The rise of diverse voices in comedy has enriched the genre, offering perspectives that challenge both gender and racial stereotypes.

Moms Mabley: A Pioneer of Intersectional Humour
As one of the first Black female comedians, Moms Mabley used her platform to address race, gender, and societal hypocrisy. Her humour was subversive, often cloaked in folksy wisdom that resonated with audiences while delivering sharp critiques.

Mabley famously said, "Love is like playing checkers. You have to know which man to move."

Modern Voices: Wanda Sykes and Ali Wong

Wanda Sykes, a master of observational humour, blends personal anecdotes with sharp commentary on race, sexuality, and politics. Ali Wong's Netflix specials, including Baby Cobra, use raw humour to explore motherhood, relationships, and cultural expectations, delivering punchlines that resonate across demographics.

Ali Wong on Motherhood

In Baby Cobra, Wong jokes, "I worked so hard to have this baby. I pumped every hour for a year. But the baby don't care—she's like, 'This formula tastes just as good.'" Her humour challenges the romanticization of motherhood, offering a refreshingly candid perspective.

The Role of Self-Deprecation in Women's Humour

Self-deprecating humour has often been a staple of women's comedy, both as a coping mechanism and a tool for social commentary. While some argue it reinforces stereotypes, others see it as a way to reclaim agency and critique societal pressures.

Lucille Ball: Laughing at Expectations

In I Love Lucy, Lucille Ball's comedic brilliance often revolved around her character's flaws and schemes. Her physical humour and willingness to appear foolish subverted traditional notions of femininity, paving the way for future female comedians.

Modern Perspectives: Confidence Over Critique

Today's comedians increasingly use self-deprecation not to diminish themselves but to highlight the absurdity of societal expectations. Hannah Gadsby's Nanette deconstructs the trope, arguing, "Self-deprecating humour is for people with low self-esteem, and I don't have it anymore."

Cultural Variations in Feminine Humour

Humour is deeply shaped by cultural contexts, and women across the globe use wit in ways that reflect their unique experiences.

British Dry Humour: Phoebe Waller-Bridge
The creator of Fleabag, Waller-Bridge epitomizes British wit with her blend of sarcasm and vulnerability. The series explores themes of grief, love, and self-destruction, often using humour to navigate the darkest moments.

The Fourth-Wall Break
In one episode, Waller-Bridge's character glances at the camera mid-seduction and deadpans, "This is a love story." The humour lies in its raw honesty, exposing the gap between societal ideals and personal reality.

Indian Humour: Comedy in a Patriarchal Society
Indian comedians like Aditi Mittal use humour to address taboo topics such as menstruation and sexism. Her stand-up routine, "Things They Wouldn't Let Me Say," blends sharp satire with cultural critique, pushing boundaries in a conservative society.

Mittal on Double Standards

Mittal jokes, "If a man stays with his parents, he's a good son. If a woman does, she's a failure to launch." The humour exposes the gendered expectations placed on familial roles.

The Feminine Wit in Everyday Life

Beyond the stage and screen, women's humour thrives in everyday interactions, from witty comebacks to group dynamics. Humour often becomes a tool for fostering connection and resilience.

Workplace Humour
In male-dominated industries, women often use humour to navigate challenging environments. One female engineer quipped during a meeting, "I'm not saying we need more women in leadership, but if you want this project done before the next ice age…"

Motherhood and Humour
Parenting humour is a genre unto itself, with mothers often sharing the chaos and absurdity of raising children. Instagram accounts and blogs like "Scary Mommy" use humour to build a community, reminding mothers they're not alone in their struggles.

The Evolution of Feminine Humour: From Margins to Mainstream

The journey of women's humour reflects broader social changes. Once relegated to the sidelines, feminine wit now occupies centre stage, reshaping comedy and society.

Breaking Taboos

Modern comedians address subjects once considered off-limits, from reproductive rights to sexual harassment. Shows like Broad City and The Marvelous Mrs. Maisel celebrate women's comedic genius while challenging stereotypes.

Expanding Platforms

Social media has democratized humour, allowing diverse voices to emerge. TikTok, Instagram, and YouTube have become platforms where women create and share comedic content, breaking barriers and building communities.

The Power of Feminine Wit

Women's humour is not just entertainment—it's a force for change, resilience, and connection. From suffrage parodies to Netflix specials, the feminine wit has evolved into a powerful tool for empowerment, challenging societal norms and amplifying marginalized voices.

As comedian Tina Fey wrote in Bossypants, "You can't be that kid standing at the top of the waterslide, overthinking it. You have to go down the chute." Feminine humour is about taking the plunge, daring to laugh at life's absurdities while shaping the world with wit.

CHAPTER 25: HUMOUR IN THE WORKPLACE

Laughing While Working

The workplace, for all its professional demands, is also a microcosm of human interaction—a place where laughter often becomes the unifying thread that connects diverse personalities, smooths over conflicts, and transforms mundane routines into meaningful collaborations. Humour, when applied with care and intention, is not merely a distraction; it is an essential tool for creativity, productivity, and connection.

This expanded chapter delves deeper into the dynamics of workplace humour, enriched with further examples, cultural insights, and stories that illuminate how laughter can elevate the professional environment.

Humour's Scientific Edge: Why Laughter Works

At its core, humour is a social signal—a way for individuals to establish trust, diffuse tension, and build rapport. In the workplace, these benefits translate into tangible outcomes, such as increased morale, reduced stress, and enhanced collaboration.

The Brain on Humour

When we laugh, our brain releases dopamine, the "feel-good" neurotransmitter, which improves focus and motivation. Simultaneously, laughter reduces cortisol, the stress hormone, creating a sense of emotional balance. For teams under pressure, these effects are invaluable.

Study Insight: Humour and Team Performance

A 2017 study in Human Relations revealed that teams with leaders who used humour appropriately during meetings experienced a 25% increase in idea generation compared to those without humour. The study concluded that humour fosters psychological safety, encouraging employees to take creative risks.

Real-Life Stories: Humour in Leadership

Barack Obama's Self-Deprecation

Barack Obama, known for his wit, often used humour to connect with his team and the public. During the White House Correspondents' Dinner, he quipped, "I may not be the young Muslim socialist I used to be." The joke, delivered with a knowing smile, diffused political tension and humanized his leadership.

The CEO's Meeting Opener

A tech CEO started a high-stakes presentation with, "Before we dive into the numbers, let me assure you —I didn't cook the books. I just microwaved them."

The humour lightened the mood and encouraged open dialogue about the company's financials.

Humour's Role in Team Dynamics: Strengthening Bonds

Teams are made up of individuals with different personalities, working styles, and stressors. Humour bridges these differences, creating a sense of shared identity.

The Power of Inside Jokes
Inside jokes are a hallmark of strong team culture. They act as shorthand for shared experiences, reinforcing bonds even during challenging times.

The "Paperclip Awards"
At an advertising agency, a team facing frequent deadlines started awarding a "Golden Paperclip" to the person who made the most creative excuse for missing a deadline. The running joke fostered camaraderie, turning frustration into laughter.

Humour Across Cultures: Navigating Global Workplaces

In today's globalized work environment, humour becomes both a bridge and a potential minefield. Understanding cultural nuances is essential for humour to connect rather than alienate.

India: Humour as Subtle Defiance
In Indian workplaces, humour often serves to challenge hierarchy while maintaining respect. Employees might joke about the boss's endless meetings, saying, "The only thing longer than his emails is his speech!" The humour,

while pointed, remains playful and inclusive.

Nordic Humour: Equality in Laughter

In Scandinavian countries, humour reflects egalitarian values. A Danish manager might tease their own work habits, saying, "I'm so efficient, I'll need another vacation to recover!" The humour breaks down hierarchical barriers, fostering a collaborative atmosphere.

American Boldness: Celebrating Wins

American workplace humour often celebrates achievement and individuality. During a company milestone, a manager might say, "We're the best team ever—don't worry, I'll add it to my LinkedIn endorsements." The humour acknowledges success while reinforcing team pride.

The Humour-Resilience Connection: Laughing Through Stress

Workplace stress is inevitable, but humour can transform even the most difficult moments into opportunities for resilience and growth.

Gallows Humour in High-Stress Jobs

In high-pressure professions like healthcare, law enforcement, and emergency response, "gallows humour" often emerges as a coping mechanism. This dark humour acknowledges the gravity of the work while providing emotional release.

The ER Quip

An ER nurse, after a chaotic shift, joked, "At least no one came in complaining about WebMD this time." The humour diffused the team's collective exhaustion, reminding them of their shared purpose.

The "Fire Drill" Fiasco

During a fire drill at a publishing house, an employee said, "Good thing we're a book company—at least we can throw something on the fire!" The joke lightened the mood during a stressful evacuation, turning chaos into camaraderie.

Humour in Conflict Resolution: Laughter as a Diffuser

Workplace conflicts can escalate quickly, but humour can de-escalate tension and shift focus toward solutions.

The Mediating Manager

A manager addressing a disagreement over resource allocation said, "Don't worry, there's enough pie for everyone. It's just that some of us might need to settle for crumbs!" The humour reframed the issue, encouraging compromise.

The Apology with a Smile

After missing a deadline, an employee sent an email with the subject line, "Late But Great (or at Least Okay)." The humour softened the apology, paving the way for constructive feedback.

Humour and Inclusion: Building Bridges

In diverse workplaces, humour can either unite or alienate. Inclusive humour celebrates differences while avoiding stereotypes or divisive topics.

Celebrating Cultural Diversity

During a global team meeting, participants shared idioms from their native languages. A German colleague shared, "Ich verstehe nur Bahnhof," which translates to "I only understand train station" (meaning "I don't understand a thing"). The laughter that followed broke down cultural barriers, creating a shared moment of joy.

Playful Gender Dynamics
A female engineer in a male-dominated team joked, "Don't worry, I'll mansplain it to you later." The humour called out gender dynamics without creating discomfort, sparking a productive conversation about inclusivity.

The Digital Shift: Humour in Remote Work

With remote and hybrid work becoming the norm, humour has adapted to virtual spaces, keeping teams connected across distances.

Virtual Icebreakers
During Zoom meetings, humorous icebreakers like "Show us your funniest coffee mug" or "Describe your day in three emojis" lighten the mood and foster connection.

The Rise of Workplace Memes
Remote teams often create memes about shared challenges, such as slow internet or endless video calls. One popular meme shows a cat sitting at a desk with the caption, "When the Wi-Fi crashes during a critical meeting." The humour turns frustration into a bonding moment.

Humour as the Secret Sauce

Humour in the workplace is more than just a break from the grind—it's a catalyst for creativity, connection, and resilience. It transforms teams into communities, leaders into relatable mentors, and challenges into opportunities for growth. When wielded with care, humour fosters a culture where employees feel valued, motivated, and inspired.

As comedian Steve Martin wisely observed, "You can't be sad while laughing." In the workplace, laughter is more than an emotion—it's a strategy for success.

CHAPTER 26: MEMES, GIFS, AND GIGGLES

Humour In The Digital World

The internet has not merely transformed humour; it has created an entire ecosystem where comedy is crowd-sourced, instantaneous, and global. In this space, memes and GIFs are more than just throwaway entertainment—they are cultural artifacts, symbols of shared experiences, and tools for shaping societal narratives. Humour in the digital age is dynamic, reflective of the zeitgeist, and often hilariously self-aware.

This deeper analysis explores not only the forms of digital humour but also their implications: how they shape cultural norms, connect us across borders, and even divide us in their wake.

The Democratization of Laughter: Everyone's a Comedian

Before the internet, comedy had gatekeepers. Stand-up stages, publishing houses, and studios controlled who got to make us laugh. Today, anyone with a smartphone and a clever idea can go viral. This democratization has created a platform for diverse voices to emerge and for humour to reach every corner of the globe.

From Viral Nobodies to Global Icons

Digital humour has propelled ordinary individuals into the spotlight overnight. A single meme or video can turn an unknown into a cultural phenomenon.

Example: Brittany Broski (The Kombucha Girl)

Brittany Broski's TikTok video of her tasting kombucha for the first time captured a spectrum of emotions—curiosity, disgust, and hesitant approval—all in a few seconds. The relatability of her exaggerated reactions turned her into an internet sensation, showcasing how humour can transform a mundane moment into universal comedy.

Memes: The Currency of Digital Culture

Memes are the bread and butter of online humour. These compact, endlessly adaptable pieces of content condense complex emotions, situations, or societal critiques into easily shareable formats.

The Evolution of a Meme

Memes rarely remain static. They mutate as they spread, with each iteration adding new layers of humour, commentary, or absurdity. Take, for instance, the "Distracted Boyfriend" meme. While it started as a simple joke about romantic disloyalty, it evolved to critique capitalism, politics, and even meme culture itself.

Example: The "Distracted Boyfriend" Meme

- Original Version: Boyfriend looks at another woman while his girlfriend glares.
- Political Version: The boyfriend is labelled "Voters," the girlfriend as "Democracy," and the other woman as "Authoritarianism."
- Meta Version: The boyfriend becomes "The Internet," the girlfriend is "Original Memes," and the other woman is "Recycled Content."

GIFs: Motion with Emotion

GIFs—those looping snippets of video—have become a universal language for expressing reactions online. Whether it's a slow clap, an eye roll, or a celebratory dance, GIFs capture emotions in ways that text cannot.

The Psychological Impact of GIFs
GIFs are effective because they tap into our love for micro-expressions and body language. A single animated gesture, like Michael Scott from The Office shouting "No!" can evoke shared understanding and laughter.

Example: "It's Happening!" GIF
This GIF, featuring Steve Carell frantically announcing, "It's happening!" during a chaotic scene, is often used to express excitement or panic. Its versatility has made it a staple of internet humour.

The Power of Participation: Remix Culture

One of the defining traits of digital humour is its participatory nature. The internet isn't just a place to consume jokes; it's a place to build on them. Memes are remixed, GIFs are captioned, and jokes are given new life in countless iterations.

Case Study: "Shrek is Love, Shrek is Life"
This absurdist meme began as a satirical video and evolved into an ironic cult phenomenon. Fans created everything from "Shrek memes" to entire Reddit communities celebrating the character, illustrating how digital humour thrives on collective creativity.

Global Humour: A Borderless Playground

The internet has turned humour into a global conversation, where jokes can cross borders faster than ever before. Yet, cultural differences remain a key factor in how humour is understood and appreciated.

Cultural Universality vs. Context
While some memes and GIFs tap into universal emotions, others rely heavily on cultural context. For example:
- Universal Humour: A GIF of a toddler falling while learning to walk elicits laughter across cultures because it's relatable.
- Culturally Specific Humour: A meme referencing Japan's Hanami (cherry blossom viewing) may not resonate with someone unfamiliar with the tradition.

Example: The Global Appeal of "Doge"
The Doge meme, featuring a Shiba Inu with captions like "Such wow. Very amaze. Much funny," transcended language barriers with its absurd simplicity. Its universal appeal lies in its playful tone and visual absurdity.

Humour as a Mirror of Society

Digital humour often serves as a lens for examining societal issues, offering biting commentary on politics, culture, and human behaviour.

Political Memes: Activism Through Humour

During the 2020 U.S. presidential election, memes became a vehicle for political commentary. The infamous "Fly on Mike Pence's Head" meme during a vice-presidential debate spawned endless iterations, with captions ranging from "This fly knows more about policy than him" to "2020 summed up in one image."

Social Commentary

Memes like "Late-Stage Capitalism" use humour to critique societal inequities. For instance, a meme juxtaposing an extravagant billionaire yacht with a crumbling public school highlights economic disparities in a way that is both humorous and thought-provoking.

The Dark Side of Digital Humour

While humour can unite and entertain, it also has a shadow side. Internet humour can perpetuate stereotypes, normalise harmful behaviour, or become a tool for trolling.

Trolling and Meme Warfare

Humour has been weaponized in online spaces to harass individuals or spread disinformation. "Doomer" memes, for instance, started as a critique of nihilism but were co-opted by extremist groups to propagate toxic ideologies.

The Ethical Dilemma of Offensive Humour

Memes often blur the line between edgy and offensive.

A meme intended as satire can easily be misinterpreted, leading to backlash or harm.

Example: The Pepe the Frog Controversy
Originally a harmless comic character, Pepe the Frog was co-opted by extremist groups, turning what began as an innocent meme into a symbol of hate. Its transformation highlights how humour can be misused in digital spaces.

Coping Through Laughter: Humour in Crisis

In moments of global or personal crisis, digital humour becomes a coping mechanism, helping people process collective trauma.

Pandemic Memes
The COVID-19 pandemic saw a surge in memes that turned fear and uncertainty into moments of levity. From jokes about toilet paper hoarding to memes about "quarantine haircuts," humour became a way to find solidarity in shared struggles.

Example: "My Plans vs. 2020"
This viral meme format used side-by-side images to contrast optimistic pre-pandemic expectations with the grim reality of 2020. It resonated universally, turning collective disappointment into collective laughter.

The Future of Digital Humour: Emerging Trends

As technology evolves, so will humour. Artificial intelligence, virtual reality, and augmented reality are already shaping the next wave of comedic content.

AI-Generated Comedy

AI tools like DALL·E and ChatGPT are now generating humorous content, from creating absurd images to writing jokes. While AI humour is still in its infancy, it raises questions about the future of creativity.

Example: AI Meme Generators

Websites like "This Meme Does Not Exist" allow users to generate random memes using AI, often resulting in bizarre but hilarious combinations.

Virtual Reality and Humour

In VR platforms like AltspaceVR, users are experimenting with immersive comedy, such as virtual stand-up shows where avatars perform for digital audiences.

The Giggles That Connect Us

Digital humour is more than just entertainment—it's a cultural force that shapes how we see ourselves and each other. From memes that critique societal norms to GIFs that capture universal emotions, online humour reminds us that laughter is both a deeply personal and profoundly communal act. In a world that often feels divided, digital humour serves as a bridge, connecting us through the shared joy of a clever punchline or an absurd meme.

As meme culture teaches us, "If you can't laugh at it, you're probably not on the internet." In the digital age, humour is the glue that holds the online world together.

CHAPTER 27: ARTIFICIAL INTELLIGENCE AND HUMOUR

Can Machines Be Funny?

Humour is a uniquely human trait, woven from our ability to recognize patterns, subvert expectations, and find joy in the absurd. Teaching artificial intelligence (AI) to understand and create humour is both a fascinating challenge and a test of how far machines can approximate the complexities of human cognition. While AI has shown glimpses of comedic promise, it often falters, reminding us that humour is as much about emotional nuance as linguistic cleverness.

In this expanded exploration, we'll delve deeper into how AI approaches humour, the technological strides being made, and the broader implications of machines that attempt to make us laugh.

Humour: The Everest of Artificial Intelligence

If solving complex equations is AI's bread and butter, humour is its Mount Everest. Comedy involves an intricate interplay of language, emotion, timing, and cultural understanding, areas where machines often struggle.

The Three Pillars of Humour AI Struggles With

1. Subtext and Context: Humour often depends on unspoken cultural norms or shared experiences, which machines struggle to infer.
2. Timing: Delivery is everything in comedy, and AI lacks the intuition to pause or emphasize the right moments.
3. Emotional Resonance: Humour connects people through shared emotions, which machines, devoid of feelings, can only mimic.

The Pizza Joke

A chatbot once attempted humour by saying, "Why did the pizza maker stop telling jokes? They were too cheesy." The joke was structurally sound, but the flat delivery and lack of audience adaptation made it feel robotic rather than relatable.

Teaching AI Humour: The Methods

Researchers have employed various techniques to teach AI how to generate and recognize humour, each revealing as much about human comedy as it does about machine learning.

Data-Driven Comedy

AI learns humour by analysing vast datasets of jokes, stand-up routines, and humorous scripts. By identifying patterns—such as common setups and punchlines—it can attempt to recreate them.

Example: OpenAI's GPT-3
When prompted, GPT-3 can produce jokes like:
"Why don't skeletons fight each other? They don't have the guts."
While familiar, the humour lacks originality, showcasing AI's tendency to recycle rather than invent.

Machine-Generated Wordplay
AI systems trained on linguistic algorithms excel at puns and wordplay because these forms of humour rely heavily on language structure.

Case Study: Pun Generator
An AI designed to create puns produced:
"What's a pirate's favourite programming language? Rrrrruby!"
While amusing, the humour feels mechanical, missing the human flair of spontaneity.

Sentiment and Sarcasm Detection
Teaching AI to recognize and replicate sarcasm —a cornerstone of human humour—is particularly challenging. Researchers use sentiment analysis to train systems to detect when language contradicts itself or takes on a mocking tone.

AI's Sarcasm Struggles
An AI once interpreted the phrase, "Oh great, another

meeting!" as a positive statement, failing to recognize the sarcasm. The humour inherent in the phrase went over the machine's metaphorical head.

The Strange Charm of AI Humour

AI-generated humour often veers into the absurd, producing jokes that are unintentionally funny or surreal. This "robotic awkwardness" has become its own genre of comedy, celebrated for its bizarre charm.

Example: Botnik's Comedy Writing
Botnik, an AI tool that generates comedic text, once produced the line:
"My microwave has opinions about politics, but only during thunderstorms."
The randomness of the statement elicited laughter, highlighting how AI can stumble into humour by accident.

AI's Unintended Hilarity
In one experiment, an AI trained to write restaurant reviews generated:
"The pancakes were so good, I forgot my own name and started a new life as a maple syrup tycoon."
The humour lies in its over-the-top absurdity, a happy accident born of algorithmic creativity.

When AI Gets It Wrong: Humour's Human Factor

Humour is deeply tied to cultural norms and emotional intelligence, areas where AI continues to fall short.

Cultural Disconnect

Humour often relies on shared cultural references that don't translate across borders or contexts. An AI generating jokes for a global audience risks alienating users by failing to account for these nuances.

Example: Lost in Translation
An AI trained on English humour attempted to translate a pun into French, producing a nonsensical result that baffled its audience. Humour reliant on wordplay often doesn't survive linguistic shifts.

Offensive Missteps
AI's reliance on unfiltered datasets can lead to inappropriate or offensive jokes, highlighting the ethical challenges of training machines to generate humour.

Case Study: The Microsoft Tay Debacle
In 2016, Microsoft's AI chatbot Tay, designed to engage humourously with Twitter users, was quickly co-opted by trolls. Within 24 hours, Tay began producing offensive and hateful statements, revealing the vulnerabilities of machine humour when exposed to toxic content.

AI-Assisted Humour: A Collaborative Future

Rather than replacing human comedians, AI shows promise as a tool for enhancing and supporting humour creation.

Scriptwriting Assistance
AI tools like Botnik can collaborate with writers, generating ideas or punchlines that humans refine and adapt.

Example: AI-Enhanced Sitcom Scripts
A team of writers used AI-generated dialogue to craft a parody sitcom episode. While the AI produced the raw material, the writers polished it into coherent comedy, blending human creativity with machine input.

Personalized Humour
AI can tailor jokes to specific audiences by analysing user preferences and cultural context, creating bespoke comedy experiences.

Real-Life Application: Virtual Assistants
Digital assistants like Alexa and Siri increasingly incorporate humour, responding to questions like, "Tell me a joke," with quips tailored to user interests or regional trends.

Ethical Implications of AI Humour

As AI humour becomes more prevalent, ethical considerations take centre stage. How do we ensure that machine-generated jokes respect cultural sensitivities and avoid perpetuating stereotypes?

Bias in Training Data
AI's humour reflects the datasets it's trained on, which can include biases or offensive content.

Example: Filtering Problematic Jokes
A humour-generating AI trained on unmoderated internet forums produced jokes that reinforced gender and racial stereotypes. Developers had to implement stricter filters to prevent such content.

The Future of Funny Machines

As AI continues to evolve, its comedic capabilities will likely improve, opening up new possibilities for humour in entertainment, education, and beyond.

AI-Generated Performances
Imagine a stand-up comedy show where AI performs alongside human comedians, generating punchlines in real time based on audience reactions.

Humour in Virtual Reality
In virtual reality environments, AI could create immersive comedic experiences, such as interactive joke-telling games or improvised humour based on user input.

Humour in Education
AI-powered humour could make learning more engaging by creating jokes tailored to specific topics, such as math or history.

The Uncanny Valley of Laughter

AI's attempts at humour, while imperfect, reveal both the promise and the limitations of machine learning. Humour remains a deeply human skill, rooted in shared experiences, emotional nuance, and cultural understanding. As machines continue to learn, they may inch closer to true comedic capability—but for now, their awkwardness is often their funniest trait.

As comedian Robin Williams once said, "Comedy is acting out optimism." For AI, humour is more than a punchline

DR BHASKAR BORA

—it's a journey into understanding the heart of human creativity.

CHAPTER 28: THE GLOBALIZATION OF LAUGHTER

Humour In A Connected World

In an era where the internet has transformed the globe into a virtual village, humour has become a universal currency. Through platforms like Reddit, YouTube, and TikTok, laughter now transcends geographical, cultural, and linguistic boundaries. While humour has always reflected societal values and experiences, globalization has amplified its reach, enabling collaborative creativity that bridges diverse perspectives.

This chapter delves into how humour connects us across borders, exploring its power to foster understanding, reflect cultural nuances, and sometimes even misfire in translation.

The Universality of Humour: Laughter as a Shared Human Experience

Humour is a universal human trait, rooted in the

ability to recognize absurdities, celebrate creativity, and find joy in connection. Across cultures, laughter signals camaraderie and reduces tension, making it a vital tool for global interaction.

Common Threads of Humour
Certain forms of humour resonate universally, transcending cultural barriers:
- Physical Comedy: Slapstick humour, such as Charlie Chaplin's iconic routines, is universally understood because it relies on visual cues rather than language.
- Situational Relatability: Jokes about universal experiences, like dealing with technology glitches or navigating awkward social interactions, often strike a chord across cultures.

Example: The Banana Peel
The classic gag of slipping on a banana peel has elicited laughter worldwide for decades. Its enduring appeal lies in its simplicity and relatability—an unexpected misstep that surprises and amuses.

Digital Platforms: The Global Stage for Humour

The internet has revolutionized humour, creating platforms where jokes, memes, and videos can gain global traction in moments.

Reddit: The Global Joke Factory
Reddit's communities, or subreddits, are hubs for collaborative humour. Subreddits like r/funny and r/ComedyCemetery showcase a mix of cultural jokes, witty observations, and meme-worthy moments.

Example: "Meanwhile, in..." Memes

The "Meanwhile, in [country]" meme format humourously highlights cultural stereotypes, often blending absurdity with affection. A post reading, "Meanwhile, in Canada: A moose and a bear sharing a Tim Hortons coffee," reflects playful exaggerations of Canadian culture.

YouTube: Humour Without Borders
YouTube's global reach allows creators from different countries to share comedic content with diverse audiences. Channels like MrBeast, Bhuvan Bam, and Lilly Singh combine local humour with universally relatable themes.

Example: Lilly Singh's Cross-Cultural Comedy
Lilly Singh, aka "Superwoman," blends her Indian heritage with Western humour, creating skits that resonate with audiences from both cultures. Her jokes about overprotective Indian parents and Western dating norms highlight cultural differences while fostering understanding.

Memes: The Universal Language of the Internet

Memes have emerged as a lingua franca of digital humour, transcending language barriers through their reliance on visuals, symbols, and simple captions.

The Spread of Global Memes
Memes like "Distracted Boyfriend," "Mocking Spongebob," and "Woman Yelling at a Cat" have achieved global recognition, with each culture adding its own twist.

Example: The "Mocking Spongebob" Meme

This meme, showing a distorted image of Spongebob accompanied by alternating uppercase and lowercase text, became a worldwide phenomenon for mocking absurd arguments. Variations appeared in multiple languages, each adapting the humour to local debates.

Localized Humour
While memes are often universal, localization tailors them to specific audiences. A meme about coffee might feature espresso in Italy, tea in India, and matcha in Japan, reflecting regional preferences while maintaining a shared structure.

Collaborative Humour: Bridging Cultures Online

The internet has enabled collaborative humour, where users from different cultures contribute to a shared comedic narrative.

TikTok Trends
TikTok fosters global collaboration through trends that invite users to riff on a theme. Dance challenges, comedic skits, and hashtag trends often gain international traction, showcasing the universality of humour.

Example: The "Sea Shanty" Phenomenon
A Scottish TikTok user posted a rendition of a traditional sea shanty, sparking a global trend where users added harmonies, instruments, and parodies. The humour of blending old-world songs with modern creativity highlighted the platform's ability to connect diverse audiences.

Cross-Cultural Comedy
Collaborative videos on platforms like YouTube often

involve creators from different countries exploring each other's comedic traditions. These projects highlight the commonalities in humour while celebrating cultural differences.

Example: A Comedy Collaboration
In a YouTube series, comedians from the UK and India swapped jokes about their respective cuisines. While the British comedian poked fun at bland food, the Indian counterpart quipped about fiery spices, illustrating how humour can bridge cultural divides.

When Humour Misfires: The Challenges of Global Comedy

While humour can unite, it can also divide when cultural nuances are overlooked or jokes fail to translate.

Lost in Translation
Puns, wordplay, and idiomatic expressions often lose their comedic impact when translated into other languages.

Example: The Pun That Didn't Translate
An English pun about "lettuce being the head of the salad" baffled a Japanese audience unfamiliar with the wordplay. The humour, reliant on linguistic structure, was lost in translation.

Cultural Sensitivity
Humour rooted in stereotypes or insensitive themes can alienate audiences rather than connect them.

Case Study: The "Kimono Controversy"
An international comedian joking about traditional

Japanese kimonos as "bathrobes" sparked backlash online. While the intent was light-hearted, the humour failed to account for the cultural significance of the attire.

Humour as a Tool for Cross-Cultural Understanding

When done thoughtfully, humour can foster empathy and understanding, offering insights into different cultures and perspectives.

Trevor Noah

Trevor Noah, host of The Daily Show, frequently uses his global upbringing to craft jokes that bridge cultural gaps. His observations about the quirks of American culture versus his South African roots resonate widely, offering a comedic lens on globalization.

"Comedy can cross borders faster than politics ever could." — Trevor Noah.

Humour During Global Crises

In times of crisis, humour becomes a unifying force, helping people cope with uncertainty and fear.

Pandemic Laughter

The COVID-19 pandemic saw a surge in global humour as people turned to memes and videos to find solace in shared struggles.

Example: The "Zoom Fails" Meme

Memes about Zoom mishaps—like forgetting to turn off the camera during embarrassing moments—united remote workers worldwide, turning collective frustrations into moments of levity.

The Future of Globalized Humour

As technology evolves, the potential for global humour will expand further, blurring the lines between local and universal comedy.

AI in Global Humour
Artificial intelligence is beginning to play a role in creating humour tailored to diverse audiences, analysing cultural trends to craft jokes that resonate across borders.

Virtual Reality Comedy
Virtual reality platforms may soon host global comedy clubs, where performers from different countries entertain live audiences, breaking down barriers through laughter.

Laughter Without Borders

The globalization of humour reminds us of the shared humanity that underpins our differences. Whether it's through a meme that goes viral in multiple languages or a TikTok trend that unites creators from around the world, laughter transcends borders, proving that joy is a universal language.

As Mark Twain said, "The human race has only one really effective weapon, and that is laughter." In a connected world, humour becomes the bridge that links us, one punchline at a time.

CHAPTER 29: THE ETHICS OF LAUGHTER

When Humour Crosses The Line

Humour's power lies in its ability to disarm, to challenge, and to connect. But with great power comes great responsibility. As societies become more interconnected and socially conscious, the ethics of humour have moved from backstage to centre stage. What once passed as an innocuous joke can now spark global outrage, and comedians are grappling with the evolving expectations of their audiences.

This expanded chapter explores the nuances of when humour crosses the line, diving deeper into the societal, cultural, and moral challenges that humour presents in today's world.

Humour and Its Dual Nature

At its core, humour is paradoxical—it can heal wounds or inflict them, bridge divides or deepen them. This duality makes it both powerful and precarious.

The Healing Side of Humour
- Relieving Tension: Laughter provides relief in tense situations, such as political crises or personal struggles. During the Great Depression, comedians like Charlie Chaplin used humour to distract audiences from economic despair.
- Uniting Communities: Humour often serves as a coping mechanism for marginalized groups, creating solidarity through shared experiences.

Example: Jewish Humour and Resilience
Jewish communities have long used self-deprecating humour to navigate hardship. As Mel Brooks quipped, "Tragedy is when I cut my finger. Comedy is when you fall into an open sewer and die." The humour underscores the absurdity of life's struggles, providing a sense of control in uncontrollable situations.

The Harmful Side of Humour
When humour targets vulnerable groups or reinforces harmful stereotypes, its effects can be damaging.
- Stereotypes in Comedy: Relying on lazy tropes, such as depictions of women as nagging wives or ethnic minorities as caricatures, perpetuates biases.
- Bullying Disguised as Jokes: Humour can be weaponized to mock or demean, often cloaked under the guise of "just kidding."

Case Study: Roast Culture Gone Wrong
In comedy roasts, participants hurl insults at one another for entertainment. However, when jokes veer into personal attacks—such as comments on someone's appearance or identity—they can leave lasting scars, as

several public figures have later attested.

The Evolution of Audience Sensitivities

In the past, comedy existed in a bubble of live performances and niche audiences. Today, jokes are immortalized on social media, accessible to global audiences with varying cultural sensitivities.

Social Media: The Amplifier of Outrage
Platforms like Twitter and Instagram have created a digital town square where jokes are scrutinized and debated in real time. A joke that resonates with one audience can be deemed offensive by another, sparking widespread backlash.

Example: Trevor Noah's Controversial Tweet
Before becoming the host of The Daily Show, Trevor Noah faced criticism for old tweets perceived as sexist. While he later apologized, the incident highlighted how social media holds comedians accountable to ever-evolving standards.

Generational Differences in Humour
Younger generations, particularly Millennials and Gen Z, tend to favour inclusive, thoughtful humour over edgy or offensive jokes. This shift reflects broader societal changes, such as the rise of movements like MeToo and Black Lives Matter, which demand greater sensitivity and accountability.

The Rise of "Safe Humour"
Comedians like John Mulaney focus on observational humour and self-deprecation, avoiding divisive topics. His joke about his mother's obsessive holiday planning

—"My mom makes Martha Stewart look like a stoner living in a van"—is both relatable and harmless, appealing to a wide audience.

When Jokes Misfire: Real-Life Examples

The Kevin Hart Oscars Controversy

In 2018, Kevin Hart stepped down from hosting the Oscars after past homophobic tweets resurfaced. While Hart expressed regret for the jokes, the incident sparked a debate about whether comedians should be judged by contemporary standards for their past actions.

- Supporters' View: Hart had already apologized for the tweets and should not have been penalized further.
- Critics' View: As a public figure, Hart's jokes perpetuated harmful stereotypes, and the apology alone wasn't enough to address the impact.

The Kathy Griffin Trump Photo

In 2017, comedian Kathy Griffin faced intense backlash after posing with a faux severed head resembling Donald Trump. The image was meant as satire but was widely criticized as crossing the line into poor taste.

- Fallout: Griffin lost endorsements, faced an investigation by the Secret Service, and struggled to rebuild her career.
- Lesson: Even satire that targets powerful figures must consider public perception and context.

Humour's Role in Social Change

Despite its controversies, humour remains a powerful tool for addressing societal issues and sparking dialogue.

Satire as Activism

Satire has a long history of challenging authority and exposing hypocrisy. Shows like The Daily Show and Last Week Tonight use humour to tackle complex issues, from income inequality to climate change, making them accessible to wider audiences.

Example: Jon Stewart and 9/11 First Responders
In a heartfelt yet humorous monologue, Jon Stewart criticized Congress for delaying healthcare funding for 9/11 first responders. The mix of outrage and wit galvanized public support, leading to legislative action.

Comedy in Marginalized Communities
For marginalized groups, humour can be a form of resistance and empowerment.
- LGBTQ+ Comedy: Comedians like Hannah Gadsby and Cameron Esposito use humour to explore queer identity and challenge stereotypes.
- Racial Satire: Writers like Hari Kondabolu critique systemic racism through humour, as seen in his documentary The Problem with Apu.

Cancel Culture vs. Creative Freedom

Cancel culture remains a polarizing topic in comedy. While some see it as a necessary mechanism for accountability, others view it as a stifling force that limits creativity.

The Case for Accountability

- Representation Matters: Jokes targeting marginalized groups can perpetuate harm and should be held to higher standards.
- Growth and Learning: Comedians who acknowledge past mistakes and adapt their material often find new ways to connect with audiences.

The Case for Creative Freedom
- Provocation as Purpose: Comedy's role is to push boundaries and provoke thought, even at the risk of offending.
- The Fear of Over-Correction: Critics argue that excessive scrutiny leads to sanitized, risk-averse humour that loses its edge.

Example: The Dave Chappelle Debate
Chappelle's Netflix special The Closer reignited debates about cancel culture, as critics accused him of transphobia while supporters defended his right to free expression. The controversy underscored comedy's evolving role in navigating sensitive topics.

Toward an Ethical Framework for Humour

The future of comedy lies not in avoiding difficult topics but in approaching them thoughtfully. An ethical framework for humour can help comedians navigate this landscape.

Key Principles of Ethical Humour
1. Intention Matters: Jokes meant to uplift or critique systems are often more impactful than those that mock individuals.
2. Punch Up, Not Down: Target those in positions of

power, not marginalized communities.

3. Be Open to Feedback: Comedians who listen to their audiences and adapt often find new ways to stay relevant.

The Delicate Art of Laughter

The ethics of humour will continue to evolve as societies grow more interconnected and socially conscious. While the line between funny and offensive may shift, the heart of comedy—its ability to make us laugh, think, and connect—remains timeless. As audiences and comedians navigate this delicate balance, humour will continue to reflect and shape the world we live in.

"Comedy is a weapon. We can use it to attack or to heal—it's all in the punchline." — Wanda Sykes.

CHAPTER 30: A FINAL REFLECTION ON HUMOUR'S ROLE IN HUMAN SURVIVAL

The Circle of Laughter

Laughter is as old as humanity itself, echoing through the ages like a song we never forgot. It began in the primal giggles of apes and grew into the playful jest of early humans, whose laughter cracked through the silence of ancient caves. Today, that same laughter roars across continents, reverberating through stand-up shows, TikTok trends, and heartfelt moments between friends.

Humour is the unbroken thread connecting the past to the present and the present to the future. It is humanity's constant refrain: "We are here, and we endure."

The Stories That Stay With Us
Laughter has carried us through the darkest moments.

In the trenches of World War I, soldiers shared gallows humour to stave off despair. In the COVID-19 pandemic, memes like "My Plans vs. 2020" brought levity to isolation, proving that even in solitude, humour connects us.

Consider the story of comedian Tig Notaro, who walked onto a stage just hours after learning she had cancer. "Good evening, I have cancer," she began, her humour raw and real. The audience laughed, cried, and clapped—not because cancer was funny, but because laughter gave them the strength to face the truth together.

These moments remind us that humour isn't escapism—it's engagement. It allows us to face life's absurdities head-on, armed with nothing but a smile and a punchline.

Laughter's Universal Language
From Shakespeare's wordplay to African trickster tales, humour transcends time and culture. An Indian comedian might joke about strict parents, an Irish storyteller might spin a yarn about pub antics, and an American meme about Zoom fails might make them both laugh. In these moments, we are reminded that humour unites us more than it divides us.

Imagine this: somewhere right now, a child in Ghana is laughing at a story about Anansi the Spider, while a teenager in Japan chuckles at a clever pun. Though separated by oceans, their laughter is the same—a shared echo of humanity's joy.

The Timelessness of Humour

Laughter outlasts the crises that birthed it. Empires fall, pandemics pass, and yet, humour remains. It adapts, finding new forms while keeping its essence intact. It is resilience wrapped in joy, a celebration of survival itself.

As artificial intelligence tries its hand at humour, we might laugh not at its jokes, but at its earnest attempts. In a thousand years, when humans live among stars, we will still tell stories and laugh at the absurdity of existence. Humour will evolve, but it will never disappear.

A Poetic Farewell

Laughter is humanity's most enduring gift.
It is the sigh after a storm, the spark in the night,
The one constant in a world forever changing.

It is born of chaos, yet it creates calm.
It divides us into breaths, yet it unites us as one.
Laughter is not just survival—it is how we thrive.

An Invitation to Laugh

Think back to the last time you truly laughed—a laugh so pure it left your cheeks sore and your soul lighter. That laugh was your defiance against despair, your connection to those around you, and your reminder that even in the face of uncertainty, joy prevails.

Carry that laughter forward. Share it generously. For as long as we laugh, we endure. And as long as we endure, we are never alone.

Final Line:

Humanity may someday lose its words or its stars, but it will never lose its laughter. And that, perhaps, is the truest sign that we were here.

EPILOGUE: THE ECHO OF LAUGHTER

In the end, what remains of us are not the monuments we build, nor the battles we fight, but the echoes of our joy—the laughter we leave behind. Long after our words fade and our footprints vanish, it is our laughter that endures, weaving through the tapestry of time like a golden thread, unbroken and eternal.

Laughter is the purest evidence of life. It rises from the depths of despair and ascends to the heights of triumph, threading through the mundane and the miraculous alike. To laugh is to declare, boldly and unashamedly, "I am alive."

The Human Symphony

Imagine humanity as an orchestra. Each laugh is a note, unique yet harmonious, adding depth to the grand symphony of our existence. There's the soft chuckle of a parent watching their child's first steps, the hearty guffaw of friends sharing old stories, the uproarious cheer of a stadium crowd united by a shared victory.

These laughs transcend barriers. They leap across borders, shatter walls, and dissolve differences. They are the universal language, one that doesn't need translation or explanation, only an open heart.

The Weight and Lightness of Laughter

Laughter is paradoxical. It is light, fleeting, and ephemeral, yet it carries profound weight. It can diffuse tensions that words cannot, soften the blows of grief, and stitch together wounds left by time and circumstance. In laughter, we glimpse the absurdity of our struggles and the beauty of our resilience.

Take the laughter that springs from shared tragedy—what we call gallows humour. It is not mockery but defiance, a way to reclaim power in powerless moments. A nurse in an emergency room, joking after a gruelling shift, or a soldier in the trenches of war, crafting a pun to distract from the chaos—these are not trivial laughs. They are acts of survival, testaments to humanity's unyielding spirit.

The Legacy of Laughter

Laughter does not belong to any one time or place. It belongs to us all, a gift passed down through generations. The jokes of ancient philosophers, the playful banter of medieval jesters, the sharp wit of modern comedians—all contribute to a legacy that defines what it means to be human.

Historical Echoes:

- Aristophanes, in ancient Greece, using satire to challenge authority and delight audiences.
- Shakespeare's fool, speaking truths that others dare not utter, hidden beneath layers of wordplay.
- The memes and TikTok trends of today, where humour connects billions across digital landscapes.

Each laugh ripples outward, touching lives we will never meet, in times we will never see.

Laughter's Place in the Universe

If aliens were to visit Earth and ask, "What defines you as a species?" we could answer with laughter. Not our machines or monuments, not our wars or wealth, but our ability to laugh at ourselves and with one another.

In laughter, we glimpse the infinite. It is the universe marvelling at itself, a burst of joy amidst the vast, indifferent cosmos. Perhaps laughter is the closest thing we have to a universal truth—a reminder that even in a universe governed by entropy and chaos, beauty and connection persist.

A Final Thought: Carrying Laughter Forward

As you close this book, pause for a moment. Think of the laughter that has shaped your life—the belly laughs with friends, the quiet chuckles over inside jokes, the unexpected giggles that caught you off guard. Each one is a thread in the fabric of your story, binding you to the people you've loved, the places you've been, and the moments you've cherished.

Laughter is an inheritance, but it is also a gift. Share it generously, for its supply is infinite. Let it echo in your homes, your workplaces, your communities. Let it lighten the burdens of strangers and deepen the bonds with those you hold dear.

For as long as we laugh, we are never alone. Laughter is our greatest rebellion, our simplest joy, and our most profound act of hope.

Closing Reflection:
Laughter is not the absence of struggle; it is its answer. It is the space between breaths where joy resides. It is humanity at its most vulnerable, most connected, and most alive. Wherever life leads us—across continents, through crises, into the unknown—we will carry our laughter like a lantern, illuminating the path and reminding us that even in darkness, there is light.

Laughter is not just a sound; it is the soul's reminder that we were here. Carry it forward, and let the world remember.

REFERENCES

The following references were used to inform and enrich the insights presented in Why We Laugh: The Role of Humour in Human Survival. These include academic studies, books, historical accounts, and cultural artifacts that support the themes explored in this book.

Books and Academic Articles

1. Bergson, H. (1911). Laughter: An Essay on the Meaning of the Comic. London: Macmillan.
   - A foundational text exploring the mechanics and philosophy of laughter.

2. Miller, G. (2000). The Mating Mind: How Sexual Choice Shaped the Evolution of Human Nature. New York: Doubleday.
   - Discusses humour as an evolutionary trait, particularly in the context of sexual selection.

3. Koestler, A. (1964). The Act of Creation. New York: Macmillan.
   - Examines humour as a creative phenomenon, linking it to art and science.

4. Provine, R. R. (2000). Laughter: A Scientific Investigation. New York: Viking.
   - A biological and social exploration of why humans

laugh.

5. Apte, M. L. (1985). Humour and Laughter: An Anthropological Approach. Ithaca: Cornell University Press.
   - Explores cultural and anthropological perspectives on humour.

6. McGraw, P., & Warner, J. (2014). The Humour Code: A Global Search for What Makes Things Funny. New York: Simon & Schuster.
   - A humorous yet academic journey through the psychology and sociology of humour.

7. Frankl, V. E. (1946). Man's Search for Meaning. Boston: Beacon Press.
   - Insights on humour as a coping mechanism, particularly during times of profound suffering.

8. Dunbar, R. I. M. (1998). Grooming, Gossip, and the Evolution of Language. Cambridge: Harvard University Press.
   - Explores laughter's role in social bonding among humans and primates.

9. Chapman, A. J., & Foot, H. C. (Eds.). (1976). Humour and Laughter: Theory, Research, and Applications. New York: Wiley.
   - A collection of multidisciplinary studies on humour.

10. Martin, R. A. (2007). The Psychology of Humour: An Integrative Approach. Amsterdam: Elsevier.
    - A comprehensive guide to the psychological underpinnings of humour.

Scientific Studies and Reports
1. Provine, R. R. (1993). "Laughter: A Vocal Behaviour of Humour and Play." American Scientist, 81(1), 38–45.
 - A scientific breakdown of laughter's origins and functions.

2. Gervais, M., & Wilson, D. S. (2005). "The Evolution and Functions of Laughter and Humour: A Synthetic Approach." The Quarterly Review of Biology, 80(4), 395–430.
 - Discusses humour from an evolutionary perspective.

3. Cousins, N. (1976). "Anatomy of an Illness." The New England Journal of Medicine.
 - Chronicles how humour played a role in the author's recovery from a life-threatening illness.

4. Martin, R. A., & Lefcourt, H. M. (1983). "Sense of Humour as a Moderator of the Relation Between Stressors and Moods." Journal of Personality and Social Psychology, 45(6), 1313–1324.
 - Examines how humour alleviates stress and enhances well-being.

5. Bressler, E. R., & Balshine, S. (2006). "The Influence of Humour on Desirability." Evolution and Human Behaviour, 27(1), 29–39.
 - Investigates humour as an attractive trait in mate selection.

Historical and Cultural Sources
1. Aristophanes. Various works, including The Clouds and Lysistrata.

- Classic Greek comedies that reveal humour's societal and political dimensions.

2. Shakespeare, W. (1599). As You Like It.
   - Explores humour's role in human folly and social commentary.

3. Molière. (1664). Tartuffe.
   - Examines the use of humour to critique hypocrisy and authority.

4. Mark Twain. Various works, including The Adventures of Tom Sawyer.
   - Illustrates humour as a vehicle for exploring human nature.

5. Charlie Chaplin's films, including Modern Times (1936) and The Great Dictator (1940).
   - Highlights humour's power to critique societal issues and inspire resilience.

Modern Media and Digital Culture
1. Reddit Subreddits:
   - r/funny, r/comedy, r/memes – Examples of collaborative and cross-cultural humour in online communities.

2. TikTok Trends:
   - Viral challenges, including the "Sea Shanty" phenomenon, illustrating humour's global reach.

3. YouTube Creators:
   - Channels like Lilly Singh's "Superwoman" and MrBeast for showcasing humour's cross-cultural appeal.

4. Memes:

- "Distracted Boyfriend," "Mocking Spongebob," and "Woman Yelling at a Cat" as examples of universally relatable internet humour.

Philosophical and Reflective Works
1. Nietzsche, F. (1882). The Gay Science.
   - Reflects on humour as a philosophical tool for grappling with existential questions.

2. Buber, M. (1923). I and Thou.
   - Explores the relational aspect of humour as a means of connecting "I" to "Thou."

3. Wilde, O. (1891). The Importance of Being Earnest.
   - A satire that underscores humour's ability to critique societal norms with levity.

Contemporary Voices
1. Trevor Noah. (2016). Born a Crime: Stories from a South African Childhood.
   - Blends humour with poignant reflections on apartheid and resilience.

2. Hannah Gadsby. (2018). Nanette.
   - Redefines the role of humour in storytelling, emphasizing vulnerability and truth.

3. Ali Wong. Netflix specials Baby Cobra (2016) and Hard Knock Wife (2018).
   - Uses humour to address gender, motherhood, and cultural identity.

This book celebrates the timelessness and universality of humour by drawing on a rich tapestry of disciplines, eras, and voices. These references, spanning the ancient to the contemporary, the academic to the anecdotal, serve as a testament to laughter's enduring role in human survival.

ACKNOWLEDGEMENTS

Writing Why We Laugh: The Role of Humour in Human Survival has been a journey of exploration, joy, and discovery. This book would not have been possible without the support and inspiration of many individuals and communities.

First and foremost, I extend my deepest gratitude to my family, whose laughter has been a constant source of light and encouragement throughout this project. To my friends and colleagues, thank you for indulging my endless questions about humour and sharing your own stories that enriched this book.

A special thanks to the researchers, comedians, and storytellers who have shaped the study and practice of humour across generations. Their work laid the foundation for this book, and their insights have been invaluable in understanding the role of laughter in human survival.

To the readers, whose curiosity and love for laughter inspired this book, thank you. It is your shared joy and engagement with the universal act of laughter that keeps this conversation alive.

Finally, my heartfelt appreciation goes to the team at Irene Mind for their support, guidance, and belief in this project. This book is as much theirs as it is mine.

May this book bring you as much joy in reading as I found in writing it.

Dr Bhaskar Bora
Author
November 2024

COPYRIGHT INFORMATION

Copyright © 2024 by Dr Bhaskar Bora
All rights reserved.

No part of this publication may be reproduced, distributed, or transmitted in any form or by any means, including photocopying, recording, or other electronic or mechanical methods, without the prior written permission of the publisher, except in the case of brief quotations embodied in critical reviews and certain other non-commercial uses permitted by copyright law. For permission requests, please contact the publisher at the address below:

Irene Mind
Contact: bora.dr@gmail.com

This book is a work of non-fiction. Any references to real people, events, or entities are included solely for illustrative purposes and are based on publicly available information. All interpretations, analyses, and conclusions are the sole responsibility of the author.

DISCLAIMER

The information presented in Why We Laugh: The Role of Humour in Human Survival is intended for informational and educational purposes only. While every effort has been made to ensure the accuracy and reliability of the content, the author and publisher assume no responsibility for errors or omissions.

This book does not constitute professional advice or therapy. Readers are encouraged to consult relevant professionals for specific concerns related to mental health, social issues, or any other matters addressed in this book.

The views expressed herein are solely those of the author and do not necessarily represent those of the publisher or any affiliated organizations. The inclusion of examples, studies, and historical accounts is for illustrative purposes and does not imply endorsement of any specific viewpoint, individual, or entity.

The author and publisher shall not be held liable for any actions taken based on the content of this book. Humour is subjective, and interpretations may vary. Reader discretion is advised.

Thank you for embarking on this journey into the world of laughter. May it inspire thought, connection, and, most importantly, joy.

www.ingramcontent.com/pod-product-compliance
Lightning Source LLC
Chambersburg PA
CBHW031618210526
45464CB00004B/1630